A Guide to AIDS

POCKET GUIDES TO
BIOMEDICAL SCIENCES

https://www.crcpress.com/Pocket-Guides-to-Biomedical-Sciences/book-series/
CRCPOCGUITOB

The *Pocket Guides to Biomedical Sciences* series is designed to provide a concise, state-of-the-art, and authoritative coverage on topics that are of interest to undergraduate and graduate students of biomedical majors, health professionals with limited time to conduct their own searches, and the general public who are seeking for reliable, trustworthy information in biomedical fields.

A Guide to AIDS

by
Omar Bagasra, MD, PhD
Donald Gene Pace, PhD

CRC Press
Taylor & Francis Group
Boca Raton London New York

CRC Press is an imprint of the
Taylor & Francis Group, an **informa** business

CRC Press
Taylor & Francis Group
6000 Broken Sound Parkway NW, Suite 300
Boca Raton, FL 33487-2742

Printed on acid-free paper

International Standard Book Number-13: 978-1-1380-3289-7 (Paperback)

Library of Congress Cataloging-in-Publication Data

Names: Bagasra, Omar, 1948- author. | Pace, Donald Gene, 1952- author.
Title: A guide to AIDS / Omar Bagasra and Donald Gene Pace.
Description: Boca Raton, FL : CRC Press/Taylor & Francis Group, 2017.
Identifiers: LCCN 2016044584 | ISBN 9781138032897 (paperback : alk. paper) | ISBN 9781315317281 (ebook)
Subjects: | MESH: Acquired Immunodeficiency Syndrome | HIV Infections | Socioeconomic Factors | Health Knowledge, Attitudes, Practice | Handbooks
Classification: LCC RC606.6 | NLM WC 39 | DDC 616.97/92--dc23
LC record available at https://lccn.loc.gov/2016044584

**Visit the Taylor & Francis Web site at
http://www.taylorandfrancis.com**

**and the CRC Press Web site at
http://www.crcpress.com**

Printed and bound in the United States of America by
Edwards Brothers Malloy on sustainably sourced paper

*We gratefully dedicate this book to our parents, Habib Ahmed
and Amina Habib Bagasra and Thomas Glenn and Vera Burton Pace,
whose love and dedication established the foundations
of our minds and souls.*

Contents

SECTION III   What to Do and Not Do

Series Preface

Dramatic breakthroughs and nonstop discoveries have rendered biomedicine increasingly relevant to everyday life. Keeping pace with all these advances is a daunting task, even for active researchers. There is an obvious demand for succinct reviews and synthetic summaries of biomedical topics for graduate students, undergraduates, faculty, biomedical researchers, medical professionals, science policy makers, and the general public.

Recognizing this pressing need, CRC Press has established the Pocket Guides to Biomedical Science series, with the main goal to provide state of the art, authoritative reviews of far-ranging subjects in short readable formats intended for a broad audience. Volumes in this series address and integrate the principles and concepts of the natural sciences and liberal arts, especially those relating to biomedicine and human wellbeing. Future volumes will come from biochemistry, bioethics, cell biology, genetics, immunology, microbiology, molecular biology, neuroscience, oncology, parasitology, pathology, and virology, as well as from other related disciplines.

This volume, devoted to AIDS, is the first in this series. AIDS has not been cured, but once a devastating and heartbreaking disease can now be controlled. Patients can lead full and active lives. The advances expertly summarized by Dr. Omar Bagasra and Dr. Donald Gene Pace are only part of the story that is still unfolding. The goal of this book is the same as the goal for this series—to simplify, summarize, and synthesize a complex topic so that readers can be edified without being exhausted by a lengthy and prosaic review.

We welcome suggestions and recommendations from readers and members of the biomedical community for future topics in this series and experts as potential volume authors/editors.

Dongyou Liu, PhD
RCPAQAP, Sydney, Australia

Preface

HIV Is Both Preventable and Manageable

Approximately 30% of the world's 7.4 billion people are infected with a particular deadly microbial agent. This agent annually infects about 1% of all the people on our planet. However, most of the infected people may not even know that they are infected because they show no apparent symptoms of the infection. In 2010, 8.8 million new patients were added to the ranks of the diagnosed; between 1.20 and 1.45 million infected individuals died. Four years later, the world's population increased and so did the death toll from this disease. In 2014, 140,000 of the infected children passed away. As with other major diseases, most of the dead lived and died in developing countries. Thankfully, the total number of cases has been on the decline since 2005 from this infectious agent, whereas new cases have fallen since 2002 [1]. China, home to about one in five of the earth's inhabitants, has made the most dramatic progress, with approximately an 80% reduction in the mortality rate from this disease. Most notably, infection from this agent is more common in developing countries where approximately 80% of the individuals in many Asian and African countries test positive for the infection, whereas only 5%–10% of the U.S. population tests positive [2]. The hope to completely eradicate the disease has been weakened because of a number of factors, including the difficulty in developing an effective vaccine, the expensive and time-consuming diagnostic process, and the need for many months of treatment. Up to this point, we have not been talking about AIDS! We have been describing tuberculosis [3].

Thirty-five years ago, when AIDS was first recognized, it was one of the deadliest and feared diseases known to mankind, but now it has taken the shape of one of the most common chronic diseases, like tuberculosis. Tuberculosis has been with mankind for eons of time, but when it first arrived, it was devastating. Then, slowly, it became a chronic, mostly latent disease. No one really knows why. HIV/AIDS has also become a chronic disease, not because of some unknown and mysterious immune adaptation, but because of new antiretroviral agents. The majority of infected people must take these highly active antiretroviral therapy (HAART) in the form of a mixed combination (cocktails) for the rest of their lives! A very small percentage of infected individuals have a natural immunity and never develop the disease. Somehow, they control HIV infection, but how and why still remains in the realm of theories, hypotheses, and inconclusive test-tube evidence plagued by numerous contradictions. Tuberculosis is the second-most common cause of death from infectious disease; HIV/AIDS is the first.

So, what are the main reasons why HIV/AIDS has not yet become a chronic, latent disease? There are three major ones, the first being the prohibitive cost of HAART. At this moment, antiretroviral agents are out of reach for most of the infected people on the earth, the majority of whom reside in developing countries. Even though the manufacturing of most antiretroviral agents is fairly inexpensive, manufacturers are not free from avarice and commonly seek to maximize profits. Increased access to HAART is a critical step to render HIV a latent chronic disease. Second is the increasing awareness of preventive measures, which we cover extensively later. The third reason why HIV/AIDS has not yet achieved the status of a chronic, latent disease is that discovering an effective vaccine has been frustratingly elusive. In the past 35 years, scientists have attempted to channel the disease-fighting potential of both antibody-mediated and T-cell-mediated immunity against HIV, but without any significant success. Because of HAART, however, we anticipate that in the coming decades, AIDS will become ever more routinely a chronic disease, one with which people can live.

HIV is a communicable disease, and unlike many diseases that are non-communicable (i.e., heart disease, cancer and diabetes), its transmission can be controlled. We urge stepped-up emphasis on prevention, and increased funding for implementation science. Biological science has received, and should continue to receive, substantial funding, but transmission is much more than a biological matter [4]. Political leaders, social service personnel, and economists, in close coordination with the Centers for Disease Control and Prevention (CDC), can see that implementation science is more effectively implemented. Improved ways of dealing with various diseases have changed their status from ongoing epidemics to epidemics of the past. The same can be true of AIDS. The official name of the CDC includes the word prevention, which along with control can make AIDS an epidemic that the world's inhabitants simply read about but no longer experience [5,6].

In this book, we aim to provide readers with a comprehensive overview of a wide variety of topics related to HIV/AIDS in a most precise and accurate fashion. Although we trust that it will reinforce and even expand the thinking of experts in the field, our primary aim is to provide a handy guide for general readers. We hope that it will both inform and inspire, for both are needed in dealing with a plague that is at once biological and behavioral in its causation.

References

1. Schaaf HS, Garcia-Prats AJ, du Preez K, Rautenbach C, Hesseling AC. Surveillance provides insight into epidemiology and spectrum of culture-confirmed mycobacterial disease in children. *International Journal of Tuberculosis and Lung Disease*, 2016;20(9):1249–1256. doi:10.5588/ijtld.15.0051.
2. Scott C, Cavanaugh JS, Pratt R, Silk BJ, LoBue P, Moonan PK. Human tuberculosis caused by mycobacterium bovis in the United States, 2006–2013. *Clinical Infectious Diseases*, 2016;63(5):594–601.
3. Scardigli A, Caminero JA, Sotgiu G, Centis R, D'Ambrosio L, Migliori GB. Efficacy and tolerability of ethionamide versus prothionamide: A systematic. *European Respiratory Journal*, 2016;48(3):946–952. doi:10.1183/13993003.00438-2016.
4. Nathanson N. HIV/AIDS epidemic: The whole truth. Comment on J Kaiser, What does a disease deserve [Science 2015 Nov 20;350(6263):900–2]. *Science*, 2016:351(6269):133.
5. Bakshi G. Here's what you need to know about tuberculosis. *Global Citizen*, August 18, 2016. https://www.globalcitizen.org/en/content/voici-ce-que-vous-devez-savoir-%C3%A0-propos-de-la-tube/, accessed September 5, 2016.
6. Centers for Disease Control and Prevention. Tuberculosis (TB). http://www.cdc.gov/tb/publications/factsheets/statistics/tbtrends.htm, accessed on September 5, 2016.

Acknowledgments

We express our appreciation to Claflin University, Orangeburg, South Carolina, where we are both professors, for its support of our scholarly work. A full-semester sabbatical for Dr. Pace was particularly helpful. Dr. Bagasra was granted one course release time during another semester, which further assisted in the book's progress.

Mr. Muhammad Hossain contributed his artistic talents to this work. His friendship and talent are much appreciated.

To the editors at Taylor & Francis Group, we express our gratitude for the outstanding support at every stage of the publishing process.

Finally, to our wives, children, and grandchildren, we gratefully acknowledge their patience and support while we worked on yet another time-consuming project.

About the Authors

Dr. Omar Bagasra was born in the back of a wooden oxcart somewhere on the plains of India. His refugee family migrated north during the exodus of the 25 million souls who were forced to leave their ancestral homelands when the former British colony of India was being partitioned during its struggle to become independent. At least eight million of these refugees—Sikhs, Muslims, and Hindus alike—perished in this partition. Being Muslim, Omar's family settled in the new nation of Pakistan, where his father became a successful grain merchant and where 10 more brothers and sisters were born and one was adopted. In this somewhat volatile environment, Omar grew into a young man.

At the age of 16, Omar decided to study other faiths and adopted an ascetic life style—starting as a Buddhist monk, a creed where warfare is anathema. He left his parents' home in Pakistan and travelled to a monastery in Tibet and lived as a Buddhist disciple, and then moved to the northern provinces of Pakistan and Afghanistan and visited many *faqirs*. After leading the life of a *faqir* for two years, Omar found that the *scientific* understanding of nature was just as important a path to truth as the more mystical, consciousness approach of the ascetic monks.

He therefore returned to Pakistan and enrolled at the University of Karachi, Karachi, where he earned a bachelor's and a master's degree in biochemistry. "I wanted to get even higher education," he says, "but in Pakistan at that time, that was as high as I could get." So, in 1972, he flew to Chicago's O'Hare airport—carrying just a suitcase of clothing and an extra $100 in his pocket.

Omar did not know anyone in the United States, but he soon found employment in the road construction industry and he learned to speak better English—his seventh language. He then got a better job of manufacturing brake shoes for the Ford Motor supplier in Albion, Indiana, near Fort Wayne. Omar saved his wages and enrolled at the University of Louisville, Louisville, Kentucky; soon he got his first scientific job working as a lab technician at the nearby Clark County Memorial Hospital in Jeffersonville, Indiana. By 1980, Omar had earned a PhD in microbiology and immunology. He joined a group in Albany, New York, to do his postdoctoral fellowship in infectious disease and he moved to Philadelphia, when his postdoc mentor moved to the city where Omar became a junior faculty member at Hahnemann University, Philadelphia, Pennsylvania, and a citizen of the United States.

Dr. Bagasra soon decided to go to a medical school. But admission policies at that time were very restrictive for individuals born and educated overseas and the tuition was more than he could afford, so the 32-year-old Omar— never one to be confined by national borders—went to study medicine at the Universidad Autónoma in Ciudad Juarez, Mexico. After two years of study, he went to Temple University, where he completed his clinical training. Subsequently, he completed residency in anatomic pathology at Hahnemann and Temple Universities, a fellowship in Clinical Laboratory Immunology at the Saint Christopher's Hospital for Children, while serving as a full time faculty member at Hahnemann University, all in Philadelphia, Pennsylvania.

Before coming to Claflin University, Dr. Bagasra held professorships at Hahnemann University (1980–1987) and Thomas Jefferson University in Philadelphia, where he served as the director of the Molecular Retrovirology Laboratories and the section chief of molecular diagnostics of the Center for the Study of Human Viruses, as well as a professor of medicine from 1991 to 1998. Dr. Bagasra also keeps a hand in clinical work—he is currently board eligible in anatomic pathology and a diplomat of the American Board of Medical Laboratory Immunology (ABMLI) and the American Board of Forensic Examiners, and a fellow of the American College of Forensic Examiners.

Dr. Bagasra's research interests have long been associated with the study of HIV and AIDS. In fact, he has been on the trail of the virus since 1981—the year of the first scientific report. For the past several years, he has focused on trying to gain insights into the molecular pathogenesis of HIV and the role of microRNA in protection against lentiviruses. In 1998, he was the first to clearly discuss the protective role of small RNAs against retrovirus and lentivirus (HIV and molecular immunity). His unswerving dedication to his work has resulted in more than 200 scientific articles, book chapters, and books. In 1995, he was nominated for the King Faisal Award for Medicine of Saudi Arabia. Over the past few years, he has received several national and international prestigious awards and recognitions. In 2002 and 2015, he received Faculty Scholar Awards from the American Association for Cancer Research. In 2006, he was the corecipient of the South Carolina Governor's Award for Excellence in Science. From 2002 to 2006, he also served as the Council Member of the American Association of Cancer Research (MICR-AACR). Dr. Bagasra currently serves as a professor of biology and the director of the South Carolina Center for Biotechnology at Claflin University.

The Institut Pasteur's Luc Montagnier—the discoverer of the AIDS virus and 2008 Nobel Laureate—described Dr. Bagasra as "a skilful researcher ... (and) a discerning scholar who explores new ideas," observing he already had a track record for challenging conventional wisdom and being proved correct. "Every scientist now knows that a significant percentage of circulating lymphocytes are infected with HIV ... but in 1992 his findings were highly controversial, when he published his paper in *NEJM New England Journal of Medicine*."

Currently, Dr. Bagasra is working on the cause of autism spectrum disorders. He believes that certain environmental chemicals are mainly responsible for causing genetic mutations and interference in fetal brain development.

Dr. Donald Gene Pace, professor of history and Spanish at Claflin University, Orangeburg, South Carolina, is a prolific author who has held leadership roles for 24 of his 31 years of full-time employment in higher education. At a previous institution, he was the chair of the Division of Social Science and Business for 12 years. At Claflin, he has served as the chair of the Department of History and Sociology (three years), interim chair of the Department of English and Foreign Languages (1.5 years), director of international studies (4.5 years), and the dean of the School of Humanities and Social Sciences (interim dean two years, dean 1 year). For the past six years, he has also been the director of the Critical Language Institute, through which he played the leading role in bringing Arabic, Chinese, Hindi, and Japanese language instruction to Claflin. He has been the principal investigator for a SAFRA grant (Student Aid and Fiscal Responsibility Act) (Critical Language Institute, five years of a five-year grant; Strengthening the Critical Language Institute, one year of a five-year grant) and for an Andrew W. Mellon grant (Internationalization of Humanities Courses at Claflin University, one year of a three-year grant). Dr. Pace has taught a wide variety of courses in Spanish, history, and political science. He has helped to develop and teach two online courses.

He received a bachelor's degree in Spanish from Brigham Young University, Provo, Utah, and three master's degrees (one in history from Brigham Young University, and two from the University of Kentucky, Lexington, Kentucky, in political science and Hispanic literature). He also received two PhD degrees: one from The Ohio State University, Columbus, Ohio, in history, and another from the University of Kentucky in Hispanic literature. Before coming to Claflin University in 2004, Dr. Pace was twice the winner of the Sears-Roebuck Foundation Teaching Excellence and Campus Leadership Award. In 2009, he won the James E. Hunter Award for Excellence in Teaching at Claflin University. In 2014, he was a finalist for the South Carolina Governor's Professor of the Year Award.

Dr. Pace has published numerous books, book chapters, and articles in a variety of subject areas, including American history, Latin American history, world history, urban environmental policy, Spanish drama, Latin American short story, confessional literature, and global health. Among the journals in which he has published are *Nature Medicine, Indian Journal of Medical Research, Journal of Infection in Developing Countries, Applied Immunohistochemistry and Molecular Morphology, Hispanic Journal, Bulletin of the Comediantes, Romance Quarterly, The Latin Americanist, Essays in Economic and Business History, Brigham Young University Studies,* and *Journal of the West.*

In addition to his professional achievements, Dr. Pace has been extensively involved in volunteer work. Beginning in his late teen years, he served a two-year church mission to Argentina as an unpaid volunteer. This had a profound impact on his global outlook, commitment to service, linguistic abilities, leadership capacity, and desire to teach and write about things that make a difference to society. Dr. Pace speaks Spanish and Portuguese, and reads Italian. He and his wife Deone are the parents of 15 children. Like her husband, Mrs. Pace graduated from Brigham Young University and has been involved for many years in voluntary service. To date, eight of their children have graduated from college (three as valedictorians), two have earned juris doctorate degrees, one a PhD in economics, and one a master of social work degree (she is now working on her doctorate in that field). Seven of the children are currently enrolled in college coursework. One of them has nearly completed a master of human resources degree, and another is in pharmacy school. Dr. and Mrs. Pace have strongly encouraged family members to provide extensive volunteer service, and so far nine have served church missions ranging from 6 to 24 months, in Poland, Cambodia, Chile, Mexico, Peru (two), Nevada, California (Spanish speaking), and California (Hmong speaking). A tenth is currently working with Hispanics in the southwestern United States. Education and service have played major roles in the lives of Dr. and Mrs. Pace, and their family. The current volume has been written in the spirit of that same sense of service: to persuade others to make the behavioral changes, service commitments, and public policy decisions needed to improve the world that we now inhabit, a world our descendants and others will inherit.

SECTION I
The Challenge of HIV and How to Respond to It

1
What is HIV?

Aids (AIDS in Luxembourgish) [1]

1.1 Human immunodeficiency virus is a lentivirus

Human immunodeficiency virus (HIV) belongs to a group of retroviruses called lentiviruses [2]. Lentiviruses are more evolved versions of retroviruses that infect primates, including humans. The genetic structure of retroviruses is composed of two identical copies of ribonucleic acid (RNA). The presence of two identical copies, or pairs, is unique to retroviruses because all other viruses contain only one copy of their genomes. Therefore, retroviruses have two single copies of RNA. HIV is a retrovirus and, like all other viruses, it requires a live, functioning host cell. The most unique aspect of retroviruses is that they are capable of converting their single stranded RNA into deoxyribonucleic acid (DNA). For this process, they use an enzyme called reverse transcriptase (RT). Before the discovery of retroviruses, this RT process was considered impossible, and the basic dogma was that DNA transcribed into RNA, but never the other way around, and then RNA translates into proteins. HIVs are of two basic kinds: HIV-1 and HIV-2. The latter is not a major pathogen for humans; it essentially has a narrow geographic range and is found in West Africa. In this book, whenever we use HIV, we will be talking about HIV-1.

1.2 Life cycle of HIV

A single viral HIV life form is extremely small and is called a virion. The diameter of the mature virion is from 132 nm to 146 nm. Mammalian cells, such as HeLa cells, are 2,000 mm in diameter, more than 15,000 times larger than the diameter of an HIV virion. As shown in Figure 7.1, its central portion, or core, is called the capsid. It contains two single strands of HIV RNA and the enzymes needed for HIV replication, including reverse transcriptase, protease, ribonuclease, and integrase. HIV, like all other retroviruses, has reverse transcriptase. This enzyme is not found in any of the host cells that they use to replicate. The viral core is made of the viral capsule protein p24, which surrounds two single strands of HIV RNA and the enzymes. The virus has nine genes: three genes, namely gag, pol, and env, contain the information needed to make structural proteins for new virions. The core is surrounded by a lipid bilayer that actually belongs to the host plasma membrane, and HIV virions acquire this when exiting from an infected cell. However, before they exit the host via the plasma membrane, they insert two kinds of viral glycoproteins in the host membrane: gp120 and gp41. These two glycoproteins make up env, and these protrude from the virion. These glycoproteins allow the virions to enter the next host cells. A single HIV can produce up to 10,000 new virions in 36 hours in a single host cell. In order for HIV to infect the next cell, it must first use gp120 to attach to a CD4 receptor [2–3].

1.3 Receptors

The CD4 receptor is found on CD4+ T-cells and macrophages. However, a mere attachment of HIV to CD4 molecules is not enough for the virus to gain entry to the host cells. It requires a second docking area on the CD4 cell surface. The target cells must contain one of the two coreceptors, CCR5 or CXCR4. These are chemokines. The former chemokine CCR5 is found on macrophages, and CXCR4 is found on CD4+ T-cells. Approximately 90% of all HIV infections prefer macrophages and are called the M-tropic HIV strain. Those that prefer CXCR4 are the T-tropic HIV strain.

After the HIV envelope successfully attaches to the host CD4 molecule, it becomes bound to one of the chemokine coreceptors, and then the envelope leverages a structural change within the gp41 envelope protein in order to fuse itself with the cell membrane. This enables the HIV virion to penetrate the membrane portion of the CD4. Once inside the cell, the elusive HIV virus is rendered safe from antibody attacks, although it is still vulnerable to opposition from the CD8 cells (cytotoxic T-lymphocytes or CTLs) [4].

1.4 Tropism

HIV tropism involves the type of cell that the virus attacks and infects, and in which it subsequently multiplies. HIV infects mainly the CD4+ cells (macrophages and CD4+ T-cells), but also, to a lesser degree, monocytes and dendritic cells, which are also CD4+ cells. Once infected, the cells' evolutionary function in the human body is compromised, and it becomes a veritable HIV replication factory because HIV infections involve two most important cells—macrophages and CD4+ T-cells—that are designed to initiate the adaptive immune response and to regulate the adoptive system, respectively. The resultant harm is very significant and damaging.

In order to comprehend how antiretroviral drugs function, and in particular how highly active antiretroviral therapy (HAART) works, we will now briefly describe HIV's life cycle. The simplistic view of this cycle begins with the first step: attachment of the intruding virion to CD4+ cells. The gp 120 molecule plays a key role in this. This is a very high affinity binding process, given that the affinity of gp120 and CD4 is very high. The second step involves the binding of one of the two coreceptors (CCR5 found on macrophages or CXCR4 on CD4+ T-cells). This permits the virus to enter the cell's cytoplasm. Here, the third step occurs as the virus sheds its core, and the viral enzymes take the two RNA strands and immediately convert the RNA into complementary DNA (cDNA), a process that involves reverse transcription. This process converts the RNA of the viral genome to cDNA and then carries and rapidly integrates within the genome of the host CD4+T-cell. Antiretroviral drugs commonly succeed in blocking this step in the HIV replication process. The fourth step occurs with the integration of the cDNA that has been reverse transcribed. The subsequent integration into the genomic DNA by the viral cDNA is mediated through the integrase enzyme that is virally encoded. This integrase slices a portion from the host DNA, after which it covalently links to the host DNA. At this point, viral cDNA moves into the genomic DNA. There are several antiviral drugs that are capable of inhibiting this enzyme. It is into open reading frame areas that the viral genome targets as its preferred port of entry, areas that are transcriptionally active. Following these basic steps, now fairly easy to describe but still very difficult to permanently block, the HIV virus has become integrated into the human genome in every somatic cell it has invaded. If the host cells are activated, and begin to replicate for whatever reason, the dormant DNA portion of the HIV begins to make messenger ribonucleic acid (mRNA) copies. This process, known as transcription, is a well-known means of transforming DNA into mRNA (Figure 7.1).

Consequently, the viral genes commence their journey of transcription, and viral mRNAs, produced through a host transcription process, inaugurate viral protein synthesis, which leads to viral protein synthesis, which in turn leads to closing the loop on the menacing cycle of viral replication [1]. When the virus comes to be assembled in proximity to the infected cell membrane, the viral proteins must be sliced in an exact manner before HIV's exiting one host cell en route to infecting another. Proteases, proteins that specialize in cutting enzymes, are central to this process. Many of these protease inhibitors have been identified. As the reader may have already guessed, once the HIV virus has become an integral part of the host DNA, it remains in a latent state, unable to be awakened until the host cells are somehow activated or when they begin to replicate. Once HIV has gotten a foothold in the human genome, the next best thing short of outright eradication is to keep the sleeping menace from being activated [5].

1.5 HIV resistance: CCR5/CXCR and mutations

As noted earlier, the CCR5 coreceptor, a β-chemokine receptor, is found on the surface of macrophages that are large cells commonly found when infection is present. Only some people have a mutation based on a 32-base pair deletion for the gene responsible for encoding for the receptor protein known as CCR5. Individuals who inherit this deletion from both parents are homozygous in terms of the CCR5-delta32 coreceptor. Having this mutation is generally a positive thing, but its presence is not uniform among all groups of humans. It is practically nonexistent in East Asians and native groups of the African continent. For African Americans, the level is about 1 in 50. For Caucasians, the level is much higher between approximately 10% and 15%. In terms of HIV infection, this particular genetic deletion provided significant protection because HIV infection relies on a CCR5 coreceptor that is intact and fully functional. The mutation closes an important door to HIV entry into cells. The mutation renders those who possess it and are naturally resistant. Some rare exceptions occur, for instance, when the HIV type is dual-tropic and prompts quick CD4 T-cell depletion. What if a child inherits the mutated gene from only one parent? This child would be heterozygous for CCR5 and would still have relative resistance to the HIV retrovirus. Were they to become infected, the progression of AIDS would be slower than normal [4].

An additional feature of protective capacity is associated with β-chemokine, the true quantity of chemokines found in a person's blood. At a cell's surface, these chemokines bind receptors. Therefore, the larger the number of chemokines secreted on the cellular surface, the greater the capacity to block HIV's binding power, and without this ability, HIV is prevented from entering a cell. In individuals who have been exposed to HIV but remain seronegative, and in some individuals whose AIDS severity does not progress over a long period of time (i.e., long-term nonprogressors, or LTNPs), elevated levels of chemokines with CCR5 are observed macrophage inflammatory protein [MIP-1] alpha and beta; and RANTES, the acronym for regulated on activation, normal T expressed and secreted). Among those who have displayed this resistance to HIV infection are individuals who are known to have been exposed repeatedly to HIV, including sex workers and others who have had numerous unprotected sexual encounters with infected partners [2–4].

1.6 CXCR4 coreceptor

As described earlier, the CXCR4 coreceptor, an α-chemokine, is found on the surface of CD4+ T-cells. X4 is used for its abbreviation; it is also known by the term fusin. It is used by T cell-tropic (T-tropic) strains of HIV. These strains typically appear in the latter stages of HIV infection and their proliferation parallels the phase of rapid AIDS progression. Similar to mutation in rCCR5 about 1 in 100 Caucasians has unusual resistance to HIV due to their possession to this coreceptor with its beneficial genetic defect. When infected with HIV, the body contains two HIV strains that coexist: the T-tropic and M-tropic strains. The HIV virus attaches to either one of these. As they are coreceptors, the virus is known as a dual tropic virus and is labeled R5 × 4 HIV [6].

1.7 Transmission

The HIV retrovirus is mainly transmitted in four different ways: (1) through sexual activity and its related fluids, sperm, semen, and vaginal secretions; (2) through blood, via unsanitary needles and transfusions; (3) through contact between a mother and her child during the process of child birth; and (4) through breastfeeding, although this risk is exceptionally low. Globally, heterosexual contact is the most common means through which HIV is transmitted. In the United States, this is the second most frequent manner of HIV transmission, with men who have sex with men (MSM) still the leading manner. Tremendous progress has been made in the area of HIV transmission through blood transfusion. Careful screening of donors and conscientious handling of blood products have become standard. However, the sharing of needles by intravenous drug users continues to be a serious problem. Mother-to-baby transmission, which can occur either at birth or earlier *in utero*, is estimated to happen in between 13% and 35% of pregnancies in which the mother is HIV positive. Although high HIV levels have been found in the milk from these mothers, the risk to breastfeeding infants has been very low. Fecal–oral transmission does not occur, nor is HIV spread through hugging, casual contacts, aerosols, touching the same household items, or insect bites. Health care workers should be careful to avoid needle pricks, but HIV transmission in this way has been extremely uncommon [7].

1.8 A case of HIV cure: The Berlin patient from Seattle

The fascinating story of Timothy Ray Brown, a cured HIV patient, has seemed to open up an apparent pathway to success in fighting the retrovirus, but thus far it has not proved upwardly scalable and susceptible to viability for larger number of HIV patients. Just the idea that Mr. Brown was infected but appears to have been cured has challenged investigators to rethink the advisability of spending a greater percentage of time and resources on an outright cure for AIDS. Is it simply an unrealistic dream, or does his ostensible cure merit creative interventions, and even risky experimentation including gene therapy? His recovery gave hope that maybe antiretroviral drugs are not the only viable response to AIDS. In May 1995, the now famous *Berlin patient*, in reality an American from Seattle, found out about his HIV-positive status. At the time, he was in Berlin doing translation work and after learning of the HIV-positive diagnosis given earlier to his former male partner, he decided to get tested himself. Brown disproved his partner's guess that he would live only a couple of years, and revived hopes for a cure. A German oncologist, Dr. Gero Hütter, took a creative and courageous approach in dealing with Brown's case. As the Seattle native also had acute myeloid leukemia (AML), Dr. Hütter decided to transplant stem cells from a donor who, fortunately for Brown, had inherited a gene that made him immune to HIV, homozygous CCR5delta32. The blood cells carrying this mutated form of the gene, when administered to Brown, first defeated Brown's HIV and then cured his leukemia. This was not simply a matter of coillness, a situation in which one infection benefits from another infection. It was a case of using a genetic transplant, and it worked brilliantly, twice in the same patient [3]. Before meeting his now famous patient, Dr. Hütter already knew the CCR5delta32 could grant protective benefits. He successfully looked for, and found, a potential donor with the homozygous delta32 mutation. The person was a German living in the United States who became the donor. Basically, Hütter disabled Brown's immune system, and then replaced it with a system that fit his particular circumstances. Remarkably, the therapy destroyed the HIV in Brown's body, including in the reservoirs when HIV hides and is dormant, until it later launches its attack on the body and unless protected by antiretroviral drugs. Brown does not take such medication, and does not need it. He is cured.

The hopeful news about the Seattle patient who was cured in Europe is that his cure provides cause for greater hope on several fronts. We realize that the human immune system, although only in a tiny percentage of people, can fight HIV and win. Antiretroviral drugs have proven effective in keeping HIV dormant in hideout reservoirs, but they have not been able to kill the virus outright. The successful intervention in Mr. Brown by infusing him with the mutated form of CCR5 suggests the possibility of giving other patients CCR5+ cells what have the beneficial delta32 mutations in order to promote receptors that resist HIV. This is not easy, because it involves dangerous transplantation of bone marrow; it is also very expensive. Gene therapy may offer promising cures, but thus far cost and technical issues have been major obstacles. Delta32-expressing vectors have the potential to place the gene inside CD4+ T-cells, but vector-based gene therapy is a delicate procedure that can have devastating consequences if not successful. An 18-year-old with a rare liver condition underwent gene therapy as part of a trial. He died and the University of Pennsylvania stopped gene therapy under orders from the Food and Drug Administration (FDA) [2–3]. Other incidents involving the use of retroviral gene therapy for leukemia patients have likewise had problems. The herpes simplex virus (HSV) may open further opportunities, but to date its use has not been widely accepted. Ethical concerns have also been raised including worries about heritability of genetic changes and unforeseen consequences [3,7]. This is a work in progress. Of note, there have no doubt been numerous cases of dual-tropic viruses that may already be present in many of the HIV infected people, which could impact gene therapy and possibly block the intended results [7].

A second apparent instance of cure involved the so-called *Mississippi baby*, whose highly publicized case raised hopes for further cases of cured patients. Unfortunately, the HIV virus reemerged after the infant's family had taken her off antiretroviral medications for more than two years. It is still uncertain where the HIV found reservoirs in which to hide. Repeated examinations, six to eight weeks apart, had shown the child to be seronegative, which made the documented HIV resurgence all the more painful. This case involved a mother with HIV who did not receive medication while pregnant to protect her child from infection. Following birth, an aggressive regimen of antiretroviral drugs was administered to the child. Based on the apparent success of the Mississippi baby, hopes rose that similar successes might be achieved, and a clinical test was begun to do just that. With the disappointing news, the clinical trial (still with no infants actively participating) was reevaluated. The infant now faces a long future of HAART treatments, and an awareness of a cure that did not materialize. The reversal is unfortunate, not just for the child but for AIDS researchers. HAART is expensive, and many in the world do not have adequate access to it. HIV-carrying individuals survive through HAART therapy, but do not live as long as their peers who are not HIV positive. A true cure likely would also have shifted attention and funding toward an outright cure for AIDS [8].

References

1. These sources have been helpful in finding translations for the word AIDS in various languages: In Different languages. a http://www.indifferentlanguages.com/words/aids; Translate AIDS to 115 different languages. TranslateThings.com, (accessed September 9, 2016). http://en.translatethings.com/a/i/d/aids.html, (accessed May 22, 2013).
2. Bagasra O. *HIV and Molecular Immunity: Prospect for AIDS Vaccine.* Natick, MA: Eaton Publishing, 1999.
3. Bagasra O, Pace DG. *Reassessing HIV Vaccine Design and Approaches: Towards a Paradigm Shift.* New York: Nova Scientific Publications, Inc, 2013.
4. NAM aidsmap. Receptors, co-receptors and immunity to HIV. http://www.aidsmap.com/Receptors-co-receptors-and-immunity-to-HIV/page/1391639/, (accessed July 27, 2016).
5. Sloan RD, Wainberg MA. The role of unintegrated DNA in HIV infection. *Retrovirology* 2011;8:52.
6. Hori T, Sakaida H, Sato A, Nakajima T, Shida H, Yoshie O, Uchiyama T. Detection and delineation of CXCR-4 (fusin) as an entry and fusion cofactor for T-tropic HIV-1 by three different monoclonal antibodies. *Journal of Immunology* 1998;160(1):180–186.
7. The Structure and Life Cycle of HIV. Anenberg Learner. https://www.learner.org/courses/ biology/textbook/hiv/hiv_4.html, (accessed July 29, 2016).
8. Ledford H. HIV rebound dashes hope of "Mississippi baby" cure. *Nature* 2014. http://www.nature.com/news/hiv-rebound-dashes-hope-of-mississippi-baby-cure-1.15535, (accessed July 30, 2016).

2
The Root of the Problem

Sindrom Kurang Daya Tahan Melawan Penyakit (AIDS in Malay)

2.1 Public policy and leadership

Public policy concerns require public leadership, and a narrow focus on branches rather than roots will not get to the root of public dilemmas. It may not deal effectively with the branches either. In a notable editorial in the *Wall Street Journal* (WSJ) titled "The Joy of What?" [1], the authors argued in favor of greater public openness in discussing root causes of serious social problems: "Sin isn't something that many people, including most churches, have spent much time talking about or worrying about through the years of the revolution (in moral behavior). But we will say this for sin: it at least offered a frame of reference for personal behavior. When the frame was dismantled, guilt wasn't the only thing that fell away; we also lost the guide-wire of personal responsibility." The very process of debating an issue can have a deterrent effect by persuading humans to accept the human responsibility for moral agency. Humans can choose, they can think, but they must think in a multidimensional fashion. HIV/AIDS, like other complex problems, is by nature complex. It is not simply a biological issue or a poverty issue or an economic issue. It is all of them, and much more, including a moral issue that is often related to right and wrong choices. The WSJ editorial stated that "the United States has a drug problem and a high-school-sex problem and a welfare problem and an AIDS problem … None of this will go away until more people in positions of responsibility are willing to come forward and explain, in frankly moral terms, that some of the things that people do nowadays are wrong."

2.2 Moral and religious dimension

Regardless of how AIDS has been publicly packaged, for example, as an unfortunate biological phenomenon or as an expensive burden to taxpayers, it does have religious and moral dimensions. D. Todd Christofferson, now a notable clergyman but earlier a young law clerk to the late Judge

John G. Sirica [2], who led the *Watergate* prosecution against President Richard Nixon, has expressed concern over the lack of moral guidance given to young people. He argues that youth should not be left to themselves "to understand and evaluate the alternatives that come before them." Without moral direction about right and wrong, they may be inadequately prepared to face the "vigorous, multimedia advocates of sin and selfishness." These negative forces do not simply leave youth alone to make intelligent decisions; they seek their own, often avaricious, agendas. In 2009, Christofferson lamented, "The societies in which many of us live have for more than a generation failed to foster moral discipline. They have taught that truth is relative and that everyone decides for himself or herself what is right. Concepts such as sin and wrong have been condemned as 'value judgments' [3]."

The apostle Paul, whose writings comprise much of the Christian New Testament, wrote that "the wages of sin is death" (Romans 6:23) [4]. Although his statement has strong spiritual connotations (death involves separation; sin separates one from God), it may be drawn on for wording that expresses the link between private moral decisions and suffering from biological consequences. To those who ignore common sense and common religious teachings in such common religious traditions as Islam, Judaism, and Christianity, HIV/AIDS may become a deadly problem. Societies are incapable of controlling moral, or immoral, behavior without moral teachings; indeed, many seem to have abdicated this duty and privilege. Sin, a central concept in Islam, Judaism, and Christianity, flourishes in a moral vacuum, so does HIV. Although HIV infection often occurs among guiltless individuals (including innocent newborns and blameless spouses), it is in its roots related to moral behavior that is characterized as a sin by the world's major religious traditions, and most minor traditions as well. Public policy makers, parents, teachers, and others, who influence those public moral values that guide private moral decisions, have the opportunity to influence behavioral roots, and not simply complain about menacing branches.

Rhetoric, the art of persuasion, is an art that needs to be turned toward roots rather than branches. Speaking about what has become the most influential constitution in the world, U.S. President John Adams explained: "Our constitution was made only for a moral and religious people. It is wholly inadequate to the government of any other." As proud as he was of the new nation he helped bring into being, Adams admitted that his country had "no government armed with power capable of contending with human passions unbridled by morality and religion. Avarice, ambition, revenge, or gallantry, would break the strongest cords of our Constitution as a whale goes through a net" [5]. On account of the contemporary confluence of poor human decisions and the genetic mutation of retroviruses, AIDS has become a whale that breaks one after another net of resistance. The net can be strengthened and the whale weakened, but not without moral values, the right to proclaim those values, and the courage to do so, the courage to tackle deep-rooted problems with moral persuasion, the courage to use rhetorical skills to protect the body politic.

We believe that similar to the strict laws enacted against sexual harassment, the national governments of the world should develop widespread awareness via social media regarding prevention of HIV infection. This has been widely successful in preventing Zika virus and Ebola epidemics in the developing countries. It appears that somehow we have forgotten about the seriousness of HIV/AIDS. This is still the number one killer among infectious diseases at global levels.

References

1. The Joy of What? *Wall Street Journal* A14, 1991.
2. Lives in the Law: D. Todd Christofferson,'72, Duke Law: News and Events. http://www.law.duke.edu/news/story?id=4710&u=11 (accessed October 31, 2011).
3. Christofferson DT. Moral Discipline. *Ensign.* November 2009. http://lds.org/ensign/2009/11/moral-discipline?lang=eng#footnote2-04211_000_036 (accessed October 31, 2011).
4. Bible, King James Version. http://quod.lib.umich.edu/k/kjv/browse.html (accessed November 28, 2011).
5. Charles Francis Adams. *The Works of John Adams.* Second President of the United States, 228–29 New York: Books for Libraries Press, 1969.

3

A Tale of Incomplete Hopes

Aids (AIDS in Dutch)

3.1 Early focus on antibodies

Since the first discovery of human immunodeficiency virus (HIV) as the true cause of acquired immune deficiency syndrome (AIDS), scientists have been concentrating on developing a vaccine that elicits anti-HIV antibodies [1–2]. Of course, this is the most common way that mankind has conquered almost all the infectious diseases [3]. This is the way that vaccine experts are trained during their medical and doctoral training in the laboratories. In order to prevent a deadly disease, they are taught, all one has to do is to fool the immune system into thinking that it has been attacked by a deadly microbe when, in reality, what the experts have done is to inject the person with a *killed microbe* [4]. Therefore, when we give DPT vaccine to little children to protect against diphtheria, pertussis, and tetanus, all we are doing is using a concoction that contains essentially *killed* microbes or pieces of microbes that elicit antibody responses against these microbes. The *magic* of the immune system is that once it is exposed to these microbes, proteins or antigens, it remembers their precise structures at the three-dimensional levels and immediately responds to these antigens by destroying real and live microbes, if exposed. Hence the art of vaccination was born. Therefore, we protect ourselves against mumps, measles, rubella (MMR) in this fashion and a form of a polio vaccine.

But, we ask, is this the only way of producing immunity? As a matter of fact, the real vaccine idea did not begin in this manner. It started with the realization that people who survive infection from smallpox never contract the disease again. Therefore, shamans and medical doctors alike began to *immunize* the vulnerable people by scratching the individuals with dried out scabs of *smallpox* victims. This observation led to the practice of *variolation* (the smallpox virus is called variola). A large majority of these *scratched* common people lost a temporary health battle as they suffered from localized lesions at the inoculation sites, but won the larger war as they emerged totally immune from a disease that ranks among the deadliest diseases humans have ever faced [5]. This primitive vaccine-like practice, which most probably originated in tenth-century China, initially spread to India, and had been observed in Egypt by the thirteenth century. Variolation had become widespread by the seventeenth century, and by 1721 it had spread from the Ottoman possessions to England. Edward Jenner, famous for using the nondeadly cowpox or *Vaccinia* virus (hence the word *vaccination*) to fight the very deadly smallpox virus was, himself, variolized during childhood [5]. Although effective, the actual protection mechanism involved in this protective vaccine is still not completely understood. It is widely believed that the Vaccinia virus prompts the creation of powerful antibodies that neutralize the effects of smallpox. This type of vaccination, which used a live but weakened (or attenuated) virus to provoke antibody response, has also been used in other effective vaccines, including the nonpathogenic *live* polio vaccine introduced in the 1950s by Albert Sabin and others, and the attenuated typhoid vaccine currently given orally to patients [1,6].

It is an opportune time in our discussion to introduce two kinds of immunities that are known to protect us against microbes: (1) antibody (or humoral) mediated immunity and (2) T-cell mediated immunity [7]. These immune reactions have a potent impact on viral-producing cells. Antibodies can bind to viral particles in the blood vessels or in extravascular spaces, which neutralizes them and increases their phagocytic potential [7]. When smallpox infection occurs, host cells consist of epithelial cells as well as the inner linings within cells in humans; these are locations not easily reached by humoral immunity. We highly recommend to our readers that they consider the importance of a recently discovered form of immunity, which involves intracellular microRNAs (miRNAs).

Accordingly, we will introduce our readers to a newly discovered immunity that appears to play a much bigger role in protecting us than previously imagined. This immunity is intracellular in nature and it is the most ancient of all the known immunities [1]. Our notion that intracellular miRNAs play a pivotal role in quelling smallpox replication has biological merit. At the time when the majority of immunological research was carried out (i.e., before the 1980s), miRNA-based immunity was unknown. Therefore, very little is mentioned in the scientific literature regarding the role of this immunity in protection against numerous intracellular pathogens like viruses and mycobacterium tuberculosis bacteria (TB).

In a supposedly pathbreaking article in *Science Express* in August 2011, Xueling Wu, et al. [8] described meticulously the process by which HIV neutralizing antibodies develop. Scientists from the National Institute of

Health's (NIH) Vaccine Research Center (VRC) led the investigative effort that provided what the NIH boldly announced were "vital clues to guide the design of a preventive HIV vaccine" [9]. At the heart of the research were three antibodies (Abs): VRC01, VRC03, and VRC-PG04. All three Abs were found in the blood: the first two came from a North American donor and the third from an African donor. Both donors were HIV-1 seropositive. The HIV-binding segments of the three Abs have shown remarkable potential for neutralizing the HIV-1 retrovirus. Although the three antibodies have structural similarities as well as differences, their similar regions permit the three to contribute collectively to HIV-1 neutralization by binding *to the same spot on the virus*. By doing so, the three Abs are able *to neutralize a high percentage of HIV strains from around the world* [9].

According to the director of the NIAID (National Institute of Allergy and Infectious Diseases), Anthony S. Fauci, "*This elegant research brings us another step closer to an HIV vaccine and establishes a potent new technique for evaluating the human immune response to experimental vaccines, not only for HIV, but for pathogens generally* [9]." This most recent article pursues the same path of a 2010 discovery by VRC researchers that proclaimed the HIV-neutralizing power of the Abs they were showcasing. In fact, they announced, two of the antibodies "could stop more than 90% of known global HIV strains from infecting human cells in the laboratory [9]."

The North American donor (donor 45) provided antibodies that were labeled VRC01, VRC02, and VRC03, obviously named after the VRC team who discovered the Abs and reported them publicly. This was impressive, but even more impressive was the 2011 VRC finding that they had discovered antibodies analogous to VRC01 in the blood of donors 74 and 0219, both from Africa.

3.2 The CD4 binding site

On account of the momentous discovery "that these VRC01-like antibodies all bind to the same spot on HIV in the same way," it would seem imperative that any future attempts at HIV-1 vaccine design focus heavily on including in the vaccine *a protein replica of this spot known as the CD4 binding site, to elicit antibodies as powerful as VRC01*. The constantly mutating HIV retrovirus has baffled researchers who have had considerable difficulty dealing with the perennial mutations. In contrast to the general mutational patterns of HIV, CD4 provides something predictable: a binding site that remains constant among the many HIV varieties across the planet. When seeking to infect a cell, the retrovirus consistently utilizes this CD4 binding site [9].

The VRC01-like antibodies have genes that are highly mutagenic: between 70 and 90 mutations *between the first draft that codes for a weak antibody and the final version that codes for an antibody that can neutralize HIV*. As the genes are found in the DNA of B lymphocytes, a type of immune cell, the VRC director, Gary J. Nabel, explains that vaccine research must focus on these B cells: "To make a vaccine that elicits VRC01-like antibodies, we will need to coach B cells to evolve their antibody genes along one of several pathways, which we have now identified, from infancy to a mature, HIV-fighting form [9]." Still, we remind readers that an antibody-mediated immunity against HIV has not been achieved so far in more than 400 different clinical trials.

Through creative use of a methodology known as *deep sequencing*, the Wu group tracked *the evolution of the antibody response to HIV at the genetic level*. As Peter Kwong, a coprincipal investigator of the 2011 article optimistically noted: "We found a way to read the books, or genes," in the large genetic library they had pulled together through deep sequencing, and following sophisticated analysis, were able to define the distinguishing characteristics that VRC01-like antibodies exhibit. Co-PI John R. Mascola indicated that it will now be more feasible to assess, at the preclinical or clinical levels, whether or not a proposed HIV vaccine is moving in the correct direction. "As we develop and test new HIV vaccines," he noted, "it will be possible to analyze not just antibodies in the blood, but also the specific B-cell genes that are responsible for producing antibodies against HIV [9]."

3.3 An elusive dormant virus

This type of announcement, although it sounds hopeful, tells only part of the story of fighting HIV/AIDS. Stimulating antibody reaction is not enough. HIV remains dormant; it hides within cells, and is not subject to the same antibody defenses that clear up other diseases. More than 40 species of nonhuman primates are infected with the simian immunodeficiency virus (SIV), similar to HIV in humans, and yet their infection does not progress to AIDS or other life-threatening conditions associated with immunodeficiency. We maintain that, as exciting and profitable as traditional antibodies-will-someday-defeat-AIDS research may be, traditional innate and acquired immunities simply have not succeeded in stopping AIDS. We further maintain that there is another means of defense that does stop HIV from progressing to AIDS. It is microRNA, something that can stop HIV within the cell, at a molecular level.

3.4 HIV vaccine dilemma

So far, the only HIV vaccine efficacy trial to show a small degree of success has been the RV144 trial that was conducted in Thailand in 2013. This recombinant HIV envelope glycoprotein-based vaccine resulted in a meager 31% efficacy. Surprisingly, neither broadly neutralizing antibodies nor cytolytic CD8+ T-cell responses were proven to be associated with protection against infection [10]. A few years ago, we published a comprehensive treatise documenting that actual immunity against HIV is intracellular in nature, and is due to microRNAs and antibody or CD8+ T cells do not play protective role at all [2]. Therefore, a recent article by Dr. Fauci, the director of the NIAID, supports our hypothesis; his work, backed by more than 30 years of evidence, confirms our approach [2]. However, despite these vaccine failures, more efforts are being utilized to improve on the results of RV144 in a southern African population by using multiple boosts, modified vectors, and adjuvants.

In the year 2020, results from a major HIV immunization project are expected to be available. The new study seeks to examine the effectiveness of an experimental vaccine designed specifically for the people of southern Africa, a region profoundly afflicted with AIDS. HIV subtype C is the prevalent strain in that region of Africa and it is this HIV subtype that the investigational vaccine is designed to attack. The new study, titled HVTN 702, builds on an earlier and much smaller trial, called HVTN 100, and both of these trials build on a monumental 2009 study that focused on the nation of Thailand, and involved over 16,000 Thai adults. Following the landmark trial, its sponsor, the surgeon general of the U.S. Army, proclaimed that the vaccine was not only safe for humans but had also achieved a 31% level of HIV prevention. Although far from the near universal success of vaccines that have virtually eliminated major diseases in the world, this 2009 experiment nonetheless brought tremendous hope to a research community that had consistently failed in the quest for an HIV vaccine. Given that discouraging history, the 31% level seemed encouraging. In RV144, one group took a placebo and another the vaccine regimen. Participants ranged in age from 18 to 30, and received either the placebo or a dose based on canary pox. Over a three-year period, participants in the study were tested for HIV every half year. All were HIV-free at the trial's beginning, but were also judged to be in a high-risk group. Of the approximately 8,000 individuals in the placebo group, 74 were infected by the end of the three years. This was substantially higher, and at a statistically significant level, than the 51 participants out of 8,000 that were given the experimental vaccine. The benefit was in the area of outright prevention. Those who contracted the disease did not have lower levels of HIV infection because of the vaccine. The vaccine seemed to be ineffective in reducing the severity of already present viral disease. The vaccine was found to be more effective early on in the three-month period. The 2016 study seeks to examine a different population, southern Africa, and to provide boosters to provide greater protection in a consistent manner, without the drop in effectiveness seen in the Thailand study. Vaccine production in the modern world is extraordinarily expensive. There are only four large-scale producers of vaccines in the world today, including GSK and Merck. As capitalist entities, these pharmaceutical companies focus on commercial products that can bring profits to offset costly investment. An investment of U.S. $500 million or more is needed to take a vaccine regimen through the steps from initial concept through delivery of a potential and potentially successful health care product. Because of the commercial focus, much of the serious work in vaccine development falls on governments, less wealthy biotechnology companies, or academia. These institutions, although helpful and significant, commonly lack the vast resources to carry out full-scale trials. The means of both preventing and treating HIV/AIDS have advanced so substantially that an ever-growing number of politicians, scientists, and other interested parties have developed a guarded optimism that contrasts starkly with the pessimism surrounding the topic in earlier decades. These parties maintain that by the year 2030, AIDS could be eliminated as a global pandemic, with local epidemics being stamped out much earlier. Propelling this optimistic outlook is the realization that antiretroviral (ARV) treatments not only keep AIDS from progressing but also significantly lowers the probability of transmission to others. Moreover, when ARV drugs are given to uninfected individuals, their chance of becoming infected declines significantly. Ambitious efforts to end AIDS locally by the year 2020 are being made in British Columbia, Canada, New York State, and the city that has long been a symbol of the challenges AIDS poses: San Francisco, California [11].

This type of announcement, although it sounds hopeful, tells only part of the story of fighting HIV/AIDS. Stimulating antibody reaction is not enough. HIV remains dormant; it hides within cells, and is not subject to the same antibody defenses that clear up other diseases. More than 40 species of nonhuman primates are infected with SIV, similar to HIV in humans, and yet their infection does not progress to AIDS or other life-threatening conditions associated with immunodeficiency. We maintain that, as exciting and profitable as traditional antibodies-will-someday-defeat-AIDS research may be, traditional innate and acquired immunities simply have not succeeded in stopping AIDS. We further maintain that there is another means of defense that does stop HIV from progressing to AIDS. It is microRNA, something that can stop HIV within the cell, at a molecular level.

Two of the claims from the clinical trial results that caught our eye are (1) the results that achieved 31% efficacy and (2) those that demonstrated that "neither broadly neutralizing antibodies nor cytolytic CD8+ T-cell responses were proven to be associated with protection against infection [10]." The clinical trial was carried out using 16,000 participants, and this number should immediately raise an interpretive alarm because within the human population, we have an enormous polymorphism (differences). Our 23,000 plus genes that express are both response to our varying environments and to our personalized internal clocks. Although a 31% success rate may sound encouraging, and should certainly be given a fair hearing, one might also consider the interpretation that for a vaccine, a mere 31% is not overly impressive, especially when it may be due primarily to biological differences. Each human being is genetically distinct from every other human beings, and at phenotypic levels (where outward manifestations of genetic codes are expressed) each of us expresses varying degrees of functional proteins and mRNAs, as well as different microRNAs (miRNAs, to be discussed in detail in a later section 5.6). Even the two identical twins, who not only share DNA from the same parents but also develop from one egg, at times have genetic variations, as the genes of each are subject to independent mutation. Variation between individuals, even siblings, creates differences that range from highly visible to largely hidden. Alleles (genetic alterations due to mutations) occur at different levels of frequency in human populations. So, if we look at the data closely, there were about 8,000 participants who were given placebo and other almost equal numbers were given the vaccine. All 16,000 participants were at a high risk for HIV, meaning that they may already have other sexually transmitted infectious diseases. Out of the control group that merely took a placebo, 74 became infected with HIV, whereas 51 became HIV infected from the experimental group that was vaccinated. That is 0.925% versus 0.638% [10]. The readers can decide for themselves how successful the vaccine was in fact!

In its 2015 report, How AIDS Changed Everything (http://www.unaids.org/en/resources/campaigns/HowAIDSchangedeverything), the Joint United Nations Programme on HIV/AIDS (UNAIDS) noted the exceptional progress that has occurred in preventing and treating the alarming disease during the first decade and a half of the new century. Nonetheless, it also acknowledged the obvious fact that much remains to be done. As of 2014, according to the more than 500 page report, approximately 36.9 million people are living with AIDS. A striking 70% of these dwell in one region, sub-Saharan Africa. Due to the effectiveness of antiretroviral drugs, the number of persons living with the virus continues to increase. The nation of South Africa has about 6.8 million individuals who are HIV infected, which exceeds the level of any other country. During the 2000–2014 period, the number of new infections dropped from an annual level of 3.1 million to a much lower 2 million. Meanwhile AIDS-related deaths declined to about 2 million in the earlier year to 1.2 million by 2014. Expenditures rose dramatically during the period, from U.S. $4.9 billion to over four times that amount, U.S. $1.7 billion. The expenditures are believed to have saved millions of lives, and prevented millions of new infections. An estimated 8.9 million children were spared the plight of being orphaned due to the interventions and expenditures. The ambitious goal of ending the AIDS epidemic is defined as reducing the number of AIDS deaths by 90% from current levels, and doing the same with HIV infections. The report calls for even larger expenditures, U.S. $8 and U.S. $12 billion more per year in order to achieve the interrelated goals of eliminating AIDS locally and then globally. As encouraging as the thought is that such a decline is possible, we emphasize that behavior is not always changed by increased expenditures. We maintain that heavy expenditures should be made but that a heavier emphasis must be replaced on behavioral matters. Besides the obvious need for higher levels of marital fidelity and virtuous behavior generally, practical procedures regarding testing are still very much neglected. For example, more than 70% of all adults in sub-Saharan Africa have never been tested for HIV. Moreover, in the Middle East, Eastern Europe, Central Asia, and North Africa, HIV infections have risen by more than 25% during the first 15 years of this century [12].

Controversies continue regarding public policy responses to HIV-positive individuals and the appropriate level of public funding. HIV-positive foreigners face deportation in 17 countries, and five Middle East countries keep HIV numbers down by simply forbidding these individuals to live within their borders. Most of the countries do not take such drastic positions, but all face important public policy decisions. Solid data, consistently maintained and forthrightly presented, are crucial. Political leaders look good to their constituencies when they attack problems and get results. Baseline data are needed in order to do this. Demonstrated progress can help keep the resources flowing into preventive research and care for those already infected. UNAIDS helps provide national and global data, and displays them in varied formats, including maps, data sheets, and charts [12]. The informative and engaging format used to provide easy visualization of global patterns is helpful but is limited by the data available for analysis.

3.5 Revisionist perspective

We have studied extensively the subject of making a vaccine against HIV. By way of summary, here are some highlights that may shed some light on the current ongoing efforts to produce an effective vaccine. Typically, vaccines are based on producing *protective antibodies* against a highly specific antigen(s) of a particular microbe, as in the cases of DPT, MMR, and polio. Once the antibodies are produced, the real protection remains in the memory of the immune system and the person vaccinated is protected for life! Therefore, individuals are exposed to the Zika virus, and after few weeks have antibodies in their bodies that counter this virus, these infected people will not only recover but will also be immune to this particular virus for life. In the case of HIV, the infected person develops antibodies to the virus but never becomes immune to the virus! So, a common sense conclusion would be that antibodies (as well as T-cell immunity) are not effective against this elusive virus. So, why put all the investigative eggs in one expensive basket and continue to try to elicit antibodies against HIV when this has not proven effective? In a recent article in the *Journal of the American Medical Association* (*JAMA*) titled "An HIV Vaccine Mapping Uncharted Territory," Dr. Fauci summarized his very optimistic description of a new vaccine called RV144 that is being scaled up and propped up by federal funds, although he admits that protection against infection is not provided by either cytolytic CD8+ T-cells or by broadly neutralizing antibodies [10]. To us, it is obvious that these traditional kinds of immunity have no protective effect against HIV. Thus, there is no valid reason to pursue these kinds of experiments on human beings; almost all of them are located in the poor nations and whose people are serving as veritable guinea pigs for the corporations located in wealthy nations. Testing on human subjects can be eminently useful, but unwise testing can become a social justice issue. If we want to find a valid vaccine or a preventive tool, we must be willing to navigate in uncharted waters.

References

1. Bagasra O. *HIV and Molecular Immunity: Prospect for AIDS Vaccine.* Natick, MA: Eaton Publishing, 1999.
2. Bagasra O, Pace DG. *Reassessing HIV Vaccine Design and Approaches: Towards a Paradigm Shift.* New York: Nova Scientific Publications, Inc., 2013.
3. de Kruif P. *Microbe Hunters.* Orlando, FL: Mariner Books, 2002.
4. Tomasso P. *Vacination.* Tasmania, Australia: Severed Press, 2013.

5. Hasan M, McLean E, Bagasra O. A Computational analysis to construct a potential post-exposure therapy against pox epidemic using miRNAs in silico. *Bioterror & Biodefense* 2016;7:1. http://dx.doi.org/10.4172/2157-2526.1000140).

6. Bagasra O, Pace DG. *Immunology and the Quest for an HIV Vaccine: A New Perspective*. Bloomington, IN: AuthorHouse, 2012.

7. Corey L, Gilbert PB, Tomaras GD, Haynes BF, Pantaleo G, Fauci AS. Immune correlates of vaccine protection against HIV-1 acquisition. *Science Translational Medicine* 2015; 7(310):310rv7.

8. Wu X, Zhou T, Zhu J, Zhang B, Georgiev I, Wang C, Chen X, Longo NS, Louder M, McKee K, O'Dell S, Perfetto S, Schmidt SD, Shi W, Wu L, Yang Y, Yang Z-Y, Yang Z, Zhang Z, Bonsignori M, Crump JA, Kapiga SH, Sam NE, Haynes BF, Simek M, Burton DR, Koff WC, Doria-Rose N, Connors M, NISC Comparative Sequencing Program, Mullikin JC, Nabel GJ, Roederer M, Shapiro L, Kwong PD, Mascola JR. Focused evolution of HIV-1 neutralizing antibodies revealed by crystal structures and deep sequencing. *Science Express* online on 2011;333(6049):1593–1602.

9. NIH-Led Team Maps Route for Eliciting HIV Neutralizing Antibodies: New Technique Can be Used Widely to Develop Vaccines, 2011. http://www.niaid.nih.gov/news/newsreleases/2011/Pages/HIVAntibodyEvolution.aspx (accessed August 12, 2011).

10. Fauci AS. An HIV vaccine: Mapping uncharted territory. *JAMA* 2016;316(2):143–144. doi: 10.1001/jama.2016.7538.

11. Jon Cohen. Means to an end. *Science* 2015;349(6245):226–231. doi: 10.1126/science.349.6245.226.

12. Cohen J. New report card on global HIV/AIDS epidemic. *Science* 2015. doi: 10.1126/science.aac8843.

4
To Fund or Not to Fund

4.1 White House leadership

Government funding can play a vital role in each of the three areas empha-sized by President Barack Obama: prevention, treatment, and reduction of health disparities. Standing in the East Room of the White House on July 14, 2010, President Obama outlined his administration's three-pronged approach, and made a plea for greater attention to groups that suffer disproportionately from HIV/AIDS: "We all know the statistics. Gay and bisexual men make up a small percentage of the population, but more than 50% of new infections. For African Americans, it's 13% of the population—nearly 50% of the people living with HIV/AIDS. Human immunodeficiency virus (HIV) infection rates among black women are almost 20 times what they are for white women. So, such health disparities call on us to make a greater effort as a nation to offer testing and treatment to the people who need it the most [1]." The President reminded his listeners that he real-ized "this strategy comes at a difficult time for Americans living with HIV/AIDS, because we've got cash-strapped states who are being forced to cut back on essentials, including assistance for AIDS drugs. I know the need is great. And that's why we've increased federal assistance each year that I've been in office, providing an emergency supplement this year to help people get the drugs they need, even as we pursue a national strat-egy that focuses on three central goals [1]." Consistent to his previous commitment, President Obama has helped to facilitate NIH funding of six medical centers to search for new HIV vaccines. Annual funding of U.S. $30 million per year, support for the Martin Delaney Collaboratory: Towards an HIV-1 Cure program (named after the late AIDS activist), the President's personal commitment, and support by four different institutes that are part of NIH are part of the overall commitment of his administra-tion. The funded institutions include the following: George Washington University, Washington, D.C.; the University of California, San Francisco; the Fred Hutchinson Cancer Research Center, Seattle; the Wistar Institute,

Philadelphia; the Beth Israel Deaconess Medical Center, Boston; and the University of North Carolina, Chapel Hill. This collaboratory effort complements HIV networks across the world.

Much has been discovered in the 35 years since AIDS was initially reported, and yet as far as we can tell, neither cure nor vaccine is on the horizon. Achieving a cure is particularly challenging because HIV establishes reservoirs by placing its own genetic matter into the very immune cells that the body relies for protection from pathogens, and these cells live a long time. Hiding, as it were, in these cells, HIV remains invisible to detection by the immune system and to therapies aimed at treatment. Only when actively replicating, cells can allow the integrated HIV to reproduce itself and then and only then the cytotoxic T-cell (CTL) and antibodies can target the virus. This is the main reason that HIV can hide in the plain sight because it can become part of the human DNA, which prevents ART treatments from doing their customary work, from clearing tissues and blood from cell with the latent virus, not replicating, at least at present. With the discontinuation of ART, however, the invisible virus sees its chance and launches a potentially lethal infection campaign.

The virus remains in these cells in a latent state, invisible to the immune system, and to anti-HIV therapies. ART only targets HIV when it is actively replicating, so the treatment can never clear the cells containing the latent, nonreplicating virus from an infected individual's blood and tissues. If an individual discontinues ART, the virus emerges from these cells and reconstitutes a widespread infection. Tackling this problem, therefore, requires experts with expansive knowledge of HIV pathology, genetics, and the immune system. The newly funded collaboratory activities aim to find ways to lure the latent HIV virus out of its cellular hiding places, where ART drugs can sense and destroy them. The various projects will benefit from laboratories located in five continents, including a continent that is being involved in an international investigative program aimed at curing HIV, a continent that suffered devastating AIDS tragedies but ironically never been included in such a research program before [2].

Although appreciatively acknowledging the generous anti-HIV spending, we hope that the U.S. $30 million funds consider novel and more effective ways to prevent HIV that are not unduly grounded in the hitherto ineffective focus on antibodies. As much as we applaud generous funding and presidential commitment, we also hope that an even greater emphasis on behaviors that avoid HIV will be part of the ambitious plans coming out of the nation's capital, a powerful metropolis that too often undermines institutions and practices that encourage public morality, abstinence, and marital fidelity. Defying age old moral norms and encouraging practices and partnerships that civilizations have wisely discouraged for centuries and millennia, is not only dangerous to public and private morality and risky in terms of public health, but also expensive. Administrations must discourage practices that promote HIV infection, not simply provide more money to treat problems that might have been prevented. The old maxim is true here: prevention is cheaper than cure.

4.2 Funding alone does not prevent the spread of HIV

More money certainly can help, but money is not the whole answer, nor is it the best answer. The teaching of private moral values in homes, churches, schools, and other settings need not be unduly expensive; indeed, it may be free or virtually free. A return to the basic values of moral behavior taught by the leading religious groups in the United States and world would be a cost-efficient means of tackling a public policy dilemma that even generous government support cannot tackle successfully. During this time of "cash-strapped states who are being forced to cut back on essentials, including assistance for AIDS drugs" [1], it is crucial that governments should back measure that support institutions, especially the traditional family, that can assist in fighting HIV/AIDS. Public and governmental support for measures, such as the Defense of Marriage Act, is a step in the right direction. It is important to acknowledge reported figures, examine trends, and probe for answers. Logical solutions, however, are much easier in the abstract than in real-world settings. Broken homes and broken communities surely play a part, as do discrepancies in educational opportunities, and lack of religious instruction. Poverty, health care disparities, and entrenched racial biases all contribute as well. Yet these seemingly intractable challenges may be at least as hard to change as the behaviors that cause HIV infections. The means to prevent HIV are largely known, yet because they intersect with the human freedom to choose, even the best efforts to prevent new cases have often proven as elusive as the quest for an AIDS vaccine.

4.3 Institutional capacity and responsibility

The battle against AIDS must be fought on many fronts. The biological and medical fronts are obvious and the bulk of the funding flow into those theatres of action. However, as this book stresses over and over again, fighting HIV involves promoting preventative behaviors, not simply costly coping and expenditures on cures. The world's institutions must be given preventative roles and the space in which to play those roles. For example, it is widely and correctly accepted that religions have a major impact on inspiring safe and correct behaviors. When religious freedom is weakened or crushed, whether overtly through direct discrimination or less obviously through excessive emphasis on tolerance carried to unwise extremes, these institutions are not as capable of exerting their preventative influence. Churches and other institutions, both public and private, must bear their responsibility to help with funding, but more importantly they must use relevant rhetoric, make the necessary speeches, hold the tough conversations, issue the critical warnings, and support common sense. When institutions that promote morality are undermined, more coping and money are required.

Research institutions are not always best if they are biggest. Granted large-scale research projects require adequate infrastructure; even so, pathbreaking ideas are not limited to large institutions, and neither should funding. Scalability has become a fashionable term in the world of research and implementation, and rightfully so. Ideas that can be acted out well on stages, both small and large are particularly useful. But let funding agencies, public and private, not forget the small stages. Democracy works best when power is diffused, and power is diffused best when resources are widely available. Likewise, AIDS will be most effectively countered by broadly institutionalized efforts, and broadly disseminated resources. Ultimately, the battle against HIV is acted out on numerous small stages, and educational efforts need to reach the players on these local stages. Think globally but act locally is not only helpful in ecological matters, but also in dealing with local health crises spawned by global pandemics. Capacity building is critical, but so is accountability and that accountability applies not only to the *funded* but also to the *funders*.

References

1. President Obama on the National HIV/AIDS Strategy. http://www.whitehouse.gov/photos-and-video/video/president-obama-national-hivaids-strategy#transcript, (accessed November 2, 2011).
2. NIH Expands investment in HIV cure research. National Institutes of Health (NIH). https://www.nih.gov/news-events/news-releases/nih-expands-investment-hiv-cure-research, (accessed July 20, 2016).

5

The Human Immune System May Protect against HIV

Ukimwi (AIDS in Swahili)

5.1 Immune: A key word in the HIV acronym

A key word on which the acronym HIV is based is *immune, acquired immunodeficiency virus*, meaning it destroys the body's immune system once it is left alone to ravage the body of an infected person. However, the human immune system is composed of several arms and in evolutionary terms each of these arms of the immune system has evolved at different time clocks on the evolutionary tree. Some are ancient and go back 4 billion years ago, whereas some have arrived at the 11th hour and are still newcomers. Whether old or new, all are crucial in protecting us against pathogenic agents [1,2]. A basic understanding of the immune system is essential in order to appreciate the challenge that human immunodeficiency virus (HIV) presents to our remarkable immune system. This system prevents the entry of tiny but dangerous entities that can make us sick (pathogens), and fights against them to eliminate the threat when such an invader does get inside the body. As simple and commonplace as it seems, the skin is of enormous value to humans. Dangerous viruses and bacteria, which could make us sick, commonly are stopped by this first line of defense. If these threats do enter the body, then the innate immunity as well as adaptive immunity (i.e., antibodies and T-cell mediated immunity) can form to counter them, and the adaptive immune system has a memory system to remember such invaders on later occasions, which forms the basis of vaccinations. Even at the level of tiny molecules, protection takes place. The different components of the immune system work in harmony with each other. Much is still not known about how these components work, but what we do know causes us to marvel at the complexity, and yet overall simplicity, of the multilevel protective system we call the immune system. Some further details are both informative and interesting.

5.2 Innate and adaptive immunity

What we inherit at birth is called innate. Basically, it stays as it is, and it is not changed due to its interaction with different pathogens. Once a microbe enters our body, our first line of defense is an innate immunity. These are composed of numerous pattern-recognition systems that instantly attack a microbe that is recognized as an invader. This is far different from what is called the classical (or adaptive) immune system. The two systems, innate and adaptive, work together in a concerted effort to fight against pathogens. The innate system provides a main line of defense through its phagocytic cells (neutrophil) that can recognize an invader by seeing their *patterns* on the surface and can tell when to kill them by eating them and spitting them out. They can recognize the pattern in a manner similar to the eye scanners or fingerprint recognition systems we use for security scans. They are highly successful and can get rid of the majority of the germs. For example, when we get a small cut while playing ball or shaving, a large quantity of microbes gets under our skin and into the lower levels of our body surface. They are immediately challenged by neutrophils and in a few days our cut surface shows a blister or white pus. This white pus represents the dead germs eaten up by thousands of neutrophils [2]. Their job is to initially keep infection from harming the body, or at least delaying its damage. Failure on their part to eliminate the germs causes the adaptive immunity to be activated. This system is little slower and takes days to weeks to awaken. This is a type of *buying time*, so that the adaptive immune system can mount an appropriate response to the pathogen and carry out its defensive work. Features of the body's nonspecific barriers that give basic protection are the skin, which serves as a barricade to most attacks by pathogens, and other responses, including reflexes such as sneezing or coughing, or getting rid of pathogens through secretions that leave the body. Tears flow from the eyes and help kill *germs*. Beneficial microbes called normal flora also help provide protection. These include fungi and bacteria, which remain in body cavities such as the mouth, the female vagina, and our 36-feet-long gastrointestinal pathway that extends all the way down to the anus. There is a biofilm in these areas that creates a protective environment. We generally take these features of the immune system for granted, unless we have a particular problem that shakes us from our complacency.

Innate immunity dedicates itself to rapid response; it works quickly, even within seconds, to counter an infection. Adaptive immunity, on the other hand, can take days or even month to successfully meet a challenge. Although much slower than innate immunity, adaptive immunity is remarkable for its capacity to recognize everything that it encounters, and we mean any *antigen*; it has a stunning capacity to *remember* past attacks. This archiving and recalling of the body's own history is comparable to a veteran general's ability to profit from previous experiences. Like the general, adaptive immunity gets better as time passes. Adaptive immunity deals with far more specific threats than innate immunity. Still, neither is inherently better than the other; each is remarkable, and as a team they complement and support each other. The real aim of immunity is to destroy the pathogenic microbes as soon as possible because speed is of the essence. Therefore, if a microbe is highly rapid in its invasion process, or if it has evolved to bypass the innate, adaptive, and intracellular (microRNA) defenses, the result will be the demise of the host [2]. Examples of this are Ebola or smallpox. However, this way of invasion is not highly beneficial for the invaders because this quick success results in elimination of the hosts and subsequently the removal of the very source that feeds that microbe's future reproduction [2]. The most successful invaders are those that work slowly, allowing enough time for the microbes to enter into a larger numbers of hosts without arousing the defense system, or causing any obvious damage to the host in a short time. Generally, these highly successful microbes live with the host until the host becomes old and immunologically compromised; then the adverse effects of the infection become apparent and may contribute to host's death. The examples of such microbes are tuberculosis (TB), herpes, cytomegalovirus (CMV), and candida, to mention just a few. A recent example of such an invasion would be the invasion of Iraq by the United States and its allies. It overwhelmed the Iraqi defense in an instant, but this has backfired on the United States and its allies. Now, we have chaos and millions of refugees trying to get asylum to the very nations that destroyed their stable nation from the first place. The drama in the human body also involves life and death matters, attacks and counter attacks, strategies and counterstrategies.

This cat and mouse game is generally in the human's favor as the immune system succeeds the majority of the time. The real kicker is HIV. It is an odd virus that actually targets the *control center* of the adaptive immune system! It kills the kind of T-cell called CD4+ that essentially runs and guides the operation of the adaptive immune system. A serious immune deficiency problem results that justifies the acronym AIDS.

Part of the fascinating work that the innate immune system performs is done by useful cells that used to be called scavenger cells (mostly neutrophils, eosinophils, basophils, and macrophages). These scavengers have evolutionarily learned to sense dangerous molecules, and then overwhelm them. They have evolutionary memory and not only know what is *foreign*, but also what is *perilous*. Just as not all immigrants are criminals, but some are, the scavenger cells learn to sense both foreignness and peril. Food, for example, would typically not be regarded as dangerous, even though it is foreign, similar to migrant workers who help us harvest the crops. The scavenger cells help tidy the body up, to keep it free of pathogens and other unwanted clutter. As one can guess, this is one of the oldest immune systems and has been estimated to be more than 3 billion years old.

Another key component of the immune system is called complement. This system provides the initial assistance to the innate immune system but also interacts with the adaptive, and at times even immediate response to pathogens that can multiply and spread through the body's fluids. Complement molecules are present in blood plasma.

5.3 The complement system

The complement system affords the initial assistance to neutrophils, and can independently provide an instantaneous microbe-killing response. Complement proteins coat the microbes and allow neutrophils to bind them quickly. Then, after eating them (called phagocytosis), neutrophils can destroy the microbes by digestion. The phagocytic vesicles, which look like bags full of digestive enzymes, can essentially rip the microbes into tiny pieces. Therefore, the complement system really complements the killing ability of neutrophils and the innate response to counter those pathogens capable of proliferating in bodily fluids. Complement molecules are found in blood plasma. They fasten themselves onto the surface of bacteria, and can also break through a bacterium's cell wall and membrane, rendering them leaky. This breaks or ruptures the basic structure of the bacterium, which leads to its destruction. It is similar to poking a knife through a ripe watermelon. This is termed lysis. Another approach that the complement molecule takes, as mentioned above, is to coat a particular bacterium, which prepares it to be destroyed by neutrophils and macrophages.

5.4 Phagocytic cells

The word *phago* has reference to the eating process; the word *cyto* has reference to the cells. The term *phagocytic*, therefore, signifies a process in which cells that are dangerous or not useful are eaten-up by monocytes. Phagocytic cells eat fungi, bacteria, and parasites. It is interesting to note that neutrophils are the fastest replicating cells found in our body and any time their number decrease or their function is impaired, and the human body becomes highly vulnerable to microbial infection. These cells constantly circulate in our blood and infiltrate the area of what is invaded by a microbe. Besides phagocytic cells, other cells that circulate in our blood, although at much lower levels, are called monocytes. They also migrate to the different organs in the body, and play the role of a hungry garbage collector that eats nonfunctional cells and those cells that have died.

An especially important time to have neutrophils functioning well is when a person undergoes chemotherapy, which is commonly used to check the growth of cancer cells. The genius of chemotherapy is its ability to identify and destroy proliferating cells. This is why hair falls out, and fingernails are affected; they are cells that have been identified, and then damaged or destroyed. Life is at stake if the neutrophils are threatened, as they are during chemotherapy, but thanks to modern medical practices, production in the bone marrow of neutrophils (and also the helpful macrophages) can be stimulated by medication. Chemotherapy can destroy bad (cancerous) cells, and damage to good cells can be minimized. After chemotherapy or radiation is completed, an injection of granulocytes/macrophage stimulating factor (GMCSF) is given, marketed and sold as *Neupogen*, *Filgastim*, *Sargramostim*, and *Neulasta*). This is a critically important injection that preserves both neutrophils and life itself [2].

We have passingly mentioned that neutrophils recognize so-called *foreign* microbes by pattern recognition. At the molecular and subcellular levels, this recognition is governed by Toll-like receptors or TLRs. They function as primary sensors of pathogens. This is an evolutionarily developed system. Therefore, as soon as a pathogen enters our body, this evolutionarily conserved innate immunity is activated, and recognition of pathogen-associated molecular patterns (PAMPs) on the microbe surfaces is seen. Receptors that recognize these surface patterns are called *pattern-recognition receptors*, or PRRs. TLRs are some of the key receptors in this part of the pattern-recognition system [2]. TLRs are expressed on numerous immune cells, including those that play a role in host defense, such as neutrophils, macrophages, natural killer cells, B- and T-lymphocytes, and dendritic cells. They are also expressed on the surfaces of cells that are not known to be immune cells but act in an accessory defense role, including fibroblasts, air, and gut lining cells. In any bacterial, fungal, and viral infection, the TLRs would immediately recognize the invasion and activate a cascade of immune responses, first innate and then adaptive, which would attempt to neutralize the threat. Usually, in the case of a viral attack (e.g., influenza), type 1 interferon (IFN) leads the response by neutralizing the virus through further activation of antiviral opposition. Numerous cytokines (e.g., Interleukin 1 Beta) and chemokines are secreted in the areas where the primary invasion is sensed, and the cells around the target areas become nonpermissive to the virus.

To enter the body, HIV appears to bypass the innate immune system. This evasion of PRRs is very odd and defies logic because all the known pathogens activate the innate immunity. We believe the reason that HIV can avoid the innate immune recognition is because in the evolutionary terms it is a brand new pathogen. As we have mentioned, PRRs are conserved throughout the evolution and they are more than 3 billion years old [2]. However, as we will discuss in Chapter 6, HIV is a new virus for *Homo sapiens*, and our immune system has never seen this virus before until the mid-1950s [2]. In order to find a cure for HIV, we must know how this virus came to be! And how it evolved so quickly to evade our defenses? [2].

5.5 Adaptive immune system

Adaptive immunity is intimately tied to innate immunity if the signaling system that rings the alarm is apparently not utilized [2]. However, it is unlikely that the adaptive response would completely fail, and it does not! But, it weakens because we know that when an individual is exposed to HIV it may take up to six months or more for the adaptive immunity to make antibodies. Generally, it takes only one to two weeks to mount a response against a virus such as influenza [2]. So, why such a delay? Once the antibodies are produced against HIV, their counterattack typically fairs. Acting much like a miniature Trojan horse, HIV relies on the body's immune system to attack itself from within! Generally, so helpful in fighting pathogens, classical immunity loses the battle when using traditional means to defeat the atypical HIV enemy.

Two categories of immunity, antibody-mediated (humoral immunity) and T-cell mediated immunity (CMI), and the different vaccines that rely on these types of immunity, simply have not worked. Classical immunity relies on its ability to detect *any foreign* substance, and its capacity to thwart its intent to damage the host body. This can be deadly when HIV is involved because HIV utilizes the immune system to activate, proliferate, and eventually bring death. Within the cells, a type of immunity known as *molecular immunity* utilizes exceptionally tiny strands of RNA known as miRNAs to disable viral threats. Vaccine attempts have routinely focused on classical immunity, but because these have not succeeded, it would make sense to try focusing on molecular immunity in vaccine research [2].

Classical immunity depends heavily on its use of B- and T-lymphocytes. This system, the lymphocyte-recognition system, senses that an invading substance is *foreign*, and then prompts the creation of antibodies (Abs) or CMI. When a virus or a special type of virus called a retrovirus (such as HIV), a bacteria, or some other *foreign* entity enters a person (or host), classical immunity makes its protective response. However, classical immunity, generally speaking, is limited in recognizing invaders that are outside of individual cells, or extracellular [3–5]. Although health preserving and useful in fighting extracellular pathogens, unfortunately classical immunity is not adept at opposing intracellular pathogens, or pathogens that have invaded the cellular genomes. Once a retroelement, such as HIV, is inside a cell's DNA, classical immunity is typically powerless to block its offensives. Retroviruses and other genetic parasites are able to go into the DNA and leave the human genome without even being detected through the mechanisms of classical immunity, but not so with the more discerning miRNA, which provide immune protection within the cells themselves at intracellular levels. In recent years, discoveries regarding miRNA have excited the scientific community, and raised hopes that they may hold a key to fighting HIV and other infections. Neither antibody-mediated immunity nor CMI has been able to deal effectively with *mycobacterium tuberculosis*, which is an organism that is covered by a specialized sheath that helps shield this potentially deadly organism from the protection generally afforded by classical immunity.

As tuberculosis has mutated, and drug-resistant varieties have developed, classical immunity has had even more difficulty countering this age-old menace that primarily attacks the lungs. Malaria, another long-time threat to human health, spreads through parasites that proliferate intracellularly in the red blood cells of the host. As these red blood cells usually do not have a nucleus, miRNAs are present only for a short time, because the nucleus is what produces them in the first place. Therefore, this mere absence of miRNAs, or even reduced amounts of miRNAs, makes them a good target for malarial parasites. The unfortunate result is the annual death of more than one million children, not to mention the 1.2 billion people that are sickened by *malaria*. [6]. HIV infect target cells, including the CD4+ cells that are important in classical immunity, but these dangerous viruses persist in a dormant state until they have the opportunity to divide. It seems terribly ironic that the very CD4+ cells that customarily protect against diseases are *tricked* by HIV retroviruses into helping them multiply, and eventually even cause the death of the CD4+ cells themselves. With vastly reduced numbers of these protective cells, the body can no longer fight off other illnesses, such as pneumonia, TB, Herpes, and candida with its usual effectiveness. In this way, the retrovirus HIV brings about the premature death of humans.

We maintain that classical immunity will not be able to defeat HIV because it has been unable to conquer malaria, TB, listeria, leprosy, and other intracellular-induced illnesses. Not only does classical immunity lack the ability to reliably recognize invasions by HIV viruses, but it can also hinder healing by producing antibodies that end up assisting the proliferation of deadly retroviruses. In short, classical immunity is simply not powerful to act within infected CD 4+ T-cells, and it actually plays a role in spreading HIV's destructive influence. Another dilemma that is not much discussed any more in the scientific community is the matter of enhancing antibodies [7]. First reported by Jay Levy in 1990, these anti-HIV antibodies actually help the virus spread the infection instead of killing the cells that are infected with HIV [7]. They evaluated 16 HIV-1 infected individuals and showed that seven of these 16 patient's serum (loaded with HIV-1 antibodies) increased viral infectivity and in four patients their antibodies increased infectivity to macrophages. The latter is considered a major route of HIV to the brain. They obtained sequential sera from five patients and three of these five patient's antibodies enhancing ability increased over time [7]. In three individuals, the HIV-enhancing capacity of their sera was more than 2,000 fold!

CMI counters a virus by dispatching two types of cells that mature in the thymus ([or T-cells]: CD8+ and CD4+). When these two types of T-cells travel to places of trouble, cells that are producing viruses, for instance, the T-cells can actually help to produce HIV. Fighting thus become assisting. It is as if friends become enemies, as T-cells are transformed from trusty allies to traitorous foes. CD8+ T-cells wipe out CD4+ T-cells, and previously-uninfected CD4+ T-cells promote a deadly cycle of destruction, bringing death first to immune cells and then to the infected persons themselves. This explains the vital role that miRNAs must play. What an incomplete picture of the immune system we see if miRNAs are not included. These remarkable entities can actually block HIV growth within CD4+ T cells and also in macrophages (to put simply, miRNAs afford protection that adaptive immunity cannot give, including in the fight against HIV). Recently, the astonishing potential of miRNAs has been explored, including in the battle against cancer and against myriad viruses and microorganisms. We anticipate that they will be used routinely in a variety of therapeutic regimens [8–10].

5.6 Molecular immunity

Is it now not the time to look beyond classical immunity? In the years following the onset of the acquired immune deficiency syndrome (AIDS) crisis, classical immunity appeared logical as the principal foundation for vaccine research, but in the later years of the 1990s theories of immunity emerged that focused on small dsRNA [8–10]. Creatively drawing on observations of worms and plants, this type of immunological theory about RNA interference (RNAi) looked at how RNAi can inhibit viral proliferation in plant viruses through what is known as gene silencing. Moreover, there are applications throughout eukaryotic life forms. As a result of their trailblazing work in RNAi, two U.S. scientists (Andrew Fire and Craig Mello) received the 2006 Noble Prize in physiology and medicine [2].

As a human race we know what we commonly repeat what we have experienced in life previously, and we scientists likewise generally use what we have been trained to do. Therefore, logically, researchers have used traditional, well-tested tools to analyze the emerging AIDS epidemic, but they have underestimated the severity of the challenge. Traditional investigative tools have been found to be seriously inadequate. Basing their findings on investigations of humans infected with HIV-1, as well as macaques infected with SIV, researchers have trusted in the ability of classical immunity to somehow win the immunodeficiency battle; research projects disappointingly have not been able to show that traditional immunological theories held the answers to stopping HIV progression, or curing AIDS once it had begun. Vaccines have not materialized, and many clinical trials, sometimes accompanied by the high hopes that media exposure brings, proved disappointing. Cumulative research is valuable, sometimes for what it shows to be true, and at other times for what it proves does not work. The vast research to date gives evidence that a paradigm shift is needed in HIV/AIDS research [2].

References

1. Bagasra O. *HIV and Molecular Immunity: Prospect for AIDS Vaccine*. Natick, MA: Eaton Publishing, 1999.
2. Bagasra O, Pace DG. 2013. *Reassessing HIV Vaccine Design and Approaches: Towards a Paradigm Shift*. New York: Nova Scientific Publications, Inc., 2013.
3. Schluter SF, Marchalonis JJ. Cloning of shark RAG2 and characterization of the RAG1/RAG2 gene locus. *Faseb Journal* 2003;17:470–472.
4. Helm T. Basic immunology: A primer. *Minnesota Medicine* 2004;87(5):40–44.
5. Pandrea I, Sodora DL, Silvestri G, Apetrei C. Into the wild: Simian immunodeficiency virus (SIV) infection in natural hosts. *Trends in Immunology* 2008;29:419–428.
6. Chowdhury K, Bagasra O. Edible vaccine for malaria using transgenic tomatoes of varying sizes shapes and colors to carry different antigens. *Medical Hypotheses* 2007;68:22–30.
7. Homsy J, Meyer M, Levy JA. Serum enhancement of human immunodeficiency virus (HIV) infection correlates with disease in HIV-infected individuals. *Journal of Virology* 1990;64(4):1437–1440.

8. Bagasra O, Prilliman K.P. RNA Interference: The molecular immune system. *Journal of Molecular Histology* 2004;35:545–555.

9. Bagasra O. RNAi as an anti-HIV therapy. *Expert Opinion on Biological Therapy* 2005; 5(11):1463–1474.

10. Bagasra, O, Stir AE, Pirisi-Creek L, Creek KE, Bagasra AU, Lee JS. Role of miRNAs in regulation of lentiviral latency and persistence. *Applied Immunochemistry and Molecular Morphology* 2006;14:166–169.

6
Origin of HIV

6.1 Why is it important to seek the origins of the AIDS virus?

As we have mentioned earlier, it is imperative that we know how HIV-1 came to existence. Without this knowledge we may never find the cure or vaccine for the malady. We must know why chimpanzees, our evolutionary cousins that share more than 97% of DNA with ours, can resist SIVcpz, whereas we are unable to resist a virus that originated from chimpanzees? However, if we think about it, the question is quite simple!

As human beings, we are exposed to numerous microbes that originated from chimps or other primates. For example, monkeys transmit yellow fever, falciparum malaria, Kyasanur Forest disease, tanapox, and Mayaro. These viruses then rely on mosquitoes, ticks, and biting flies to transmit disease from monkeys to humans. Simian virus 40 (SV-40) is present in macaque monkeys, and most people were exposed to it through vaccines that contained monkey tissue in the early days of Albert Sabin's live polio vaccination [1,2]. None of these diseases are as fatal as HIV. If untreated, HIV is almost 100% fatal. Why? How can a disease that is essentially innocuous to chimps be so fatal to its human hosts? Let us stress here that the notion that HIV invaded human hosts because they ate chimp meat and hunted them is so obviously ridiculous and contrived that it defies the common sense [3–5]. Man has been hunting monkeys and chimps for eons of time. Why, then, after so many years, did acquired immune deficiency syndrome (AIDS) arrived precisely at the time when the oral polio vaccine was given to more than a million people in Congo in the mid- to late-1950s? There is no physical evidence that HIV existed before this particular time period. Physical evidence for the earliest HIV sample is from 1959 (published by David Ho's team from Aaron Diamond Institute of New York [6]). In Section 6.3, we will discuss and provide scientific evidence for how HIV-1 emerged and came into humans, why it is so deadly for us but not to chimps, and why it is able to evade our highly evolved immune defense systems? Investigation into the origins of HIV-1 is critical in development of a cure for HIV-1. We do not intent to deride or blame anyone. Our goal is to honestly present the information as we see it, which we hope will lead to finding the cure for HIV-1/AIDS and to the development of a vaccine and effective therapeutics. No vaccine or cure for a deadly disease has ever come about without knowing the origin or cause of the disease (with the rare exception of smallpox).

The current scientific solution focuses heavily on providing all infected individuals with highly active antiretroviral therapy (HAART) for a life time, because the moment an HIV-positive person stops taking the prescribed anti-HIV cocktails (sometimes even missing a pill by mistake), the HIV returns right away. These treatments are so expensive that it is out of the reach of the majority of infected people. Now, the trend is to give HAART, as a preventive measure, to anyone who may be at risk of getting HIV infection! Never in human history has a medical community willingly delivered highly toxic medications to people who are not even infected. As we will explore this in the next few Sections from 7.7 through 7.9, this is the current status of medical advice. Even if HAART were free and supplies were unlimited, one must remember that antiretroviral agents are not without harm! The medical community is now recommending that newborns of HIV-infected mothers should get HAART as a preventive measure, even though there is no evidence of their infection. We must find vaccine for HIV and, until we honestly discover how HIV-1 originated, we may never find the cure for the disease.

Where did HIV come from? This has been the subject of intense investigation and a substantial global controversy [4,5]. Both of the AIDS viruses, HIV-1 and HIV-2, originated in Africa. Why does HIV evade the human immune systems (both innate and classical) when SIV in chimpanzees does not evade theirs? How did HIV adapt so rapidly to man and why it is so deadly? We present strong evidence to argue that AIDS appeared in late 1950s from the oral polio vaccine (OPV) used in clinical trials conducted in the Belgian Congo. How did a presumed big bang of zoonosis originate from chimpanzees SIVs and spread to humans when the human is not a very good host to SIVs in the first place?

The first biological samples to document what has become current human HIV-1 pandemic came from blood plasma that had, in 1959, originally been gathered for research on glucose-6-phosphate-dehydragenase deficiency and sickle cell anemia [6]. To begin with, out of more than 1,200 samples, 24 were found to be HIV positive by immunoassay methodology.

One particularly notable sample, labeled L70, was obtained in Leopoldville, Belgian Congo (presently Kinshasa, Democratic Republic of Congo) from an adult male Bantu in 1959 [6]. This sample was declared HIV-1 positive following multiple confirmatory assays. Due to its age, the specimen yielded only limited sequences, which were called Z59a, Z59b, Z59c, and Z59d. Of these Z59-short sequences, the first (a) corresponded to the region of the Envelop (*env*) where the V3 loop is located. The second (b) was located at the junction of gp120 and gp41, the parts of the virus that HIV-1 uses to attach to CD4+ cells. The third (c) corresponded to the *env* region where gp41 C-terminus (located where an amino acid chain ends; short for carboxy terminal; CDT). The fourth (d) corresponds to the *pol* (polymerase) gene region. In 1998, David Ho's research team reported these HIV-1 sequences, which provided a preliminary sense of the start of the proliferation of a major global pandemic that has killed tens of millions of people [6]. The Ho team's mapping of the spread of the virus noted that there was a distance of more than 400 miles between Kinshasa (where the M-group strain of HIV-1, the most common among humans, has been identified) and the southeastern Cameroon location of the chimpanzees with the simian form of the virus (SIV) that is most similar to the human M strains. Perhaps the Congo River provides the answer, because the virus could have spread from the location of the chimpanzees in southeastern Cameroon to the former Leopoldville (Kinshasa) in modern Democratic Republic of Congo [3–7]. But then, this interpretation may not be accurate. The Zhu team concluded, "Our results also indicate that subtypes B, D, and F may have evolved within the human population rather than arising from multiple cross-species transmission events" [6].

Many studies have linked HIV-1's origins to the hunting of infected chimpanzees, with the supposition raised that zoonosis led to a chimp-to-human leap on many occasions until, finally, it became a pandemic. We find no credible evidence that this is the case.

6.2 Strain 1, group M, and the epidemiology of HIV

HIV is a virus that evolves with exceptional rapidity. At the surface level, HIV may seem to be a single virus, but such is not even close to being the case. HIV not only has various subtypes that are genetically distinct from each other, but it is also remarkably subject to recombination. HIV-1 evolution, in particular, has plagued humanity, including the so-called *within-patient* viral adaptation. The plural term HIVs, not simply HIV, is the reality that the overall term HIV masks. Within just a few decades HIV/AIDS has gone from being unknown to assuming its place among the very most significant infectious diseases in the world, with a total number of cumulative infections now estimated at about 60 million. At the end of the year 2015, the number of individuals living with HIV was an estimated 36.7 million, of which 2.1 million were newly infected and 1.1 people died from AIDS. Although new HIV infections seem to have reached their summit in 1996, the total number of persons living with the infection has continued to rise, in part because of the success of antiretroviral drugs. Uneven rates of prevalence, and varied epidemiological patterns characterize these infections. Resistance and adaptation to antiretroviral drugs further complicate matters. Dealing with HIV is like walking on shifting sands.

Researchers who study epidemics, epidemiologists, have identified major HIV strains, groups, and subgroups. The predominant viruses are HIV-1 and HIV-2. The latter has two groups associated with it, A and B, and is endemic in the western part of the African continent. The global HIV-1 pandemic is almost entirely the result of the spread of group M (or Major Group) of the HIV-1 virus. In the epidemiological term, this is the subtype or clade and group that have been most *successful*. Epidemiologists subdivide this particularly virulent group M into nine subtypes, all of the consonants between A and K (A, B, C, D, F, G, H, J, and K). As if this were not complicated enough, more than 40 CRFs (circulating recombinant forms), which tend to have evolved more recently, add to the shifting sands of HIV complexity. The HIV/AIDS epidemic has been traced primarily to four subtypes (A, B, C, and D) and two CRFs (CRF01_AE and CRF02_AG) [4].

Newly infected persons have the greatest capacity to infect others. Unfortunately, this comes at a time when *infectors* are unaware of their infection. The reason is that in the early phase of infection (generally, about day 1 to day 10) the virus level in the blood is very high. After this, generally short phase, the viral level comes down drastically. This happens apparently in the absence of any classical immunity (i.e., antibodies or CD8 cytotoxic T-Cells responding to HIV-1 proteins). At high viral levels in the blood and semen, the infected person is very infectious and can spread HIV to other intimate partners and, if pregnant, to her unborn baby. At this phase, even if predisposed to do so, they are not likely to see a need to warn potential *infectees*. Besides, persons who spread HIV to others often do so because they act according to emotions rather than using rational common sense. The main means that have helped to fuel the current HIV epidemic are sexual activity, injection with contaminated needles, and transmission from mothers to their children while these mothers are pregnant, during the delivery of their babies. Breastfeeding is generally not considered a risk for the newborn. Transmission via sexual contact accounts for roughly 80% of the HIV infections worldwide, and most of these contacts are heterosexual.

An additional 10% of the world's HIV infections are due to injection drug use, or IDU, and the most prevalent among these are heroin and crack. Infected individuals who are responding well to antiretroviral treatments have a substantially lower probability of infecting someone else than if they were receiving no such treatments (this is simply due to very low levels of virus in their blood). Likewise, male circumcision can significantly (60%–90%) reduce the chances of a male becoming infected, which indirectly lowers female infections. The use of sanitized or clean needles, like the correct use of condoms, brings the likelihood of transmission to nearly zero [8].

What is HAART and what impact has it had? HAART stands for highly active antiretroviral therapy. Since its introduction in 1996, it has had a profound effect in lengthening the lives of HIV-infected individuals who previously were doomed to a rapid demise due to the progression of AIDS, and then death due to an AIDS-related cause. HAART uses a combination or cocktail that includes three different drugs that come from a minimum of two different classes. HAART has been so successful that AIDS has often become a chronic infection that is treatable, and that does not lead rapidly to death. Besides cost and distribution problems, which are considerable, HAART and its users have contributed to drug resistance problems. HIV, which mutates easily even without encouragement from patients who fail to adhere to prescribed routines, has undergone numerous mutations, and a serious problem known as transmitted drug resistance (TDR) has emerged [9].

The AIDS epidemic began in Africa, the Americas, and Europe before it commenced in Asia, but given Asia's huge population, and the predictions of AIDS proliferation experts predict that the world of the near future will have three Asian nations with the greatest total number of AIDS cases: China, India, and Indonesia. In China, the world's most populous country, injected drug use was initially the leading cause of HIV infection. Later the main path of infection became, and still is, sexual contact. The epidemic in this country was primarily centered in high-risk groups, but evidence now indicates that the general population has become a serious part of the problem. HIV infection rates for drug users exceed 50% in some of China's worst areas for HIV/AIDS [10]. In these highly troubled areas, women are also experiencing high prevalence rates [10,11]. More recently, China has achieved a monumental reduction in its HIV-infected population. With education, and strictly enforced laws, it has accomplished what no major nation has done so far [10,11].

In terms of the HIV/AIDS epidemic, the pattern seen in China is all too familiar. High-risk behavior in concentrated pockets of individuals initially accounts for most cases. HIV infection then becomes more generalized, and heterosexual activity in the general population comes to be a major part of the problem. China's self-imposed troubles with gender selection, and the resultant lack of females, will likely intensify the populous nation's challenges relating to AIDS. China and India, each with more than one billion people, need to control HIV infections, for their own sake and that of the entire planet. What used to be seen, with some justification, as mainly an African problem, with some serious troubles in Europe and the Americas, is no longer a justifiable position to take. It now seems reasonable to assert, and worry, that as China goes, so goes the world; and as India goes, so goes the world. These two countries are home to about one-third of the world's 7.4 billion inhabitants. HIV infection has truly become a global problem on a planet whose population is the largest it has ever been.

6.3 Origin of HIV-1

This tragic event that gave birth to the deadly AIDS pandemic seems to have emerged from an attempt to improve health and save lives through an oral polio vaccine (OPV). In those days (the 1950s) the only way to grow the attenuated, weakened poliovirus was to use monkey kidney cells harvested from live nonhuman primates. To do this, fresh monkey kidneys were macerated in sterile condition with small scissors and then grown in fresh monkey serum along with certain supplements of vitamin and other ingredients. This culture concoction seems like a witches brew to us today, but it was state-of-the-art in those early years of tissue cultures. Back then, not too many scientists were even aware of the presence of pathogenic viruses in nonhuman primates. In fact, there were not too many experiments conducted to look for any long term pathogenic effects of OPV being used at the time. If someone was very cautious, they would inject the prepared vaccines in the brain of chimpanzees to determine if an OPV had any side effects. Unfortunately, as mentioned above, all 40 plus species of African nonhuman primates are resistant to HIV or SIVs, making the experiment a nonstarter from the beginning. If OPV had been prepared in any of the African nonhuman primates' kidneys or even the serum to supplement the cell cultures, they would have allowed the growth of SIVs from various African nonhuman primates. Therefore, we believe that the OPV trial facilitated the entry of combinations of varying SIVs into single cells, which allowed their recombination and brought novel types of lentiviruses initially into the first large group of vaccines, and subsequently into human society, combinations that were capable of bypassing the regular immunological immune mechanisms [4]. Chimpanzees (*Pan troglodytes*) come in three subspecies: *Pan troglodytes troglodytes*, *Pan troglodytes verus*, and

Pan troglodytes schweinfurthii. These nonhuman primates apparently rely on an miRNA defense system that has developed over the distant past and do not develop a simian form of AIDS when infected with HIV-1 or their own SIVcpz. There are about 50 African nonhuman primates, in addition to larger primates such as gorillas and chimpanzees that carry SIVs. However, except on rare occasions, they do not suffer from any immune deficiency disease comparable to AIDS in humans. Obviously, unlocking the secret to this pattern holds great promise for fighting HIV-1 infection in humans. We maintain that to comprehend HIV-1's origin, what scientists call its pathogenesis, lies in examining the event in the former Belgian Congo that permitted multiple strains of simian (SIVs related to monkeys or apes) retroviruses (SIV) to be blended and then adapted to human intracellular defenses, which subsequently became the HIV-1 group M [3–4].

A major vaccination attempt provided the venue for what may have launched the HIV-1 epidemic, which in turn developed into a pandemic. In the normal course of human life, *Homo sapiens* would not simultaneously be exposed to the blend of lentiviruses that would infect our normally-protective CD4+ cells or, for that matter, other specific types of cells. If that were to happen, the SIVs involved would instantly begin to recombine among themselves, with the result being the emergences of untold numbers of new SIVs. Some of the new creations would be resistant to the endogenous microRNA defenses, thereby making humans vulnerable to infections against which the normal immune systems could not protect. With the passage of time, other strains of HIV-1 would emerge capable of bypassing miRNAs, with levels of homology to HIV-1 >90, which would create immunological chaos among humans. Suppose that 10 million cells were infected with various SIV strains at virtually the same time, which could occur if a vaccine utilizing the kidney cells of chimpanzees were manufactured. Suppose further that these syntheses of infections are introduced into a million people. Chances are very likely that roughly 10,000 individuals would develop variations of HIV-1, with varying levels of pathogenicity. Of these thousands of new HIV-1 variants, some could be categorized as M (major group), others N (new group), and also a large number as O (outlier group). This is what may have occurred when the OPV was given during 1957 and 1959. Included in the vast numbers of variants were many that had potential for replication. This apparently occurred, and many adults, and particularly children, survived the onslaught of disease, presumably with less virulent strains of HIV-1. Those children, now in the final years or decades of their lives, could potentially be hosts to the very subtypes needed to create an HIV-1 vaccine that is nonpathogenic and therefore safe to administer [3].

6.4 Exposure to one type of lentivirus: Pathogenic to humans, primates?

Human beings, and most other mammals, commonly come into contact with retroviruses and lentiviruses (a more complex type of retrovirus in which symptoms are delayed significantly following infection; *lenti* refers to slowness). They do this, for instance, when they butcher animals such as sheep and goats, when they touch domesticated animals such as dogs and cats, or when they come into contact with cow or pig meat. In primate colonies over a long period of time, Asian macaques have had SIV exposure, but do they become ill or acquire a life-threatening immunological deficiency? No, rarely. It is not well understood that humans, likewise, commonly neither get infected nor ill from lentiviruses and other retroviruses, including HTLV-I and -II (lymphotropic viruses that occasionally cause T-cell leukemia and other maladies), and HIV-2 (although it still can be deadly). Why is a lentivirus deadly to one individual but not to someone else? Does the answer lie in the effectiveness of the miRNA defenses that provide more protection for some persons than for others?

Although considerable attention has been given, and should continue to be, placed on HIV-1, the virus that causes AIDS, could much be learned from yet more research into HIV-2. Specifically, why are humans so affected by HIV-1 but not HIV-2? The answer to that question inform efforts to help humans develop improve immunity in fighting off, or outright preventing, the more virulent HIV-1. HIV-2 is the lentivirus that is so common in West Africa, but which typically does not cause serious illness. Humans tend to do well when only exposed to a single lentivirus, such as HIV-2. However, when faced with recombinant forms of SIVcpzs (including *schweinfurthii* and *troglodytes*), specifically those presumed to have been introduced in the Belgian Congo during the OPVs, human immunity is exceptionally vulnerable. If HIV-2 does cause serious immunological problems in humans, it is likely that they will not occur for decades after exposure, even without antiretroviral drugs. The reason for this, we believe, is that HIV-2 springs from a single SIVsm strain, not a recombinant form such as HIV-1. Most humans, with sufficient homologous miRNAs to disable HIV-2 for a long time, have not found this retrovirus nearly as challenging as HIV-1. Still, it should not be taken lightly. It is transmitted in the same ways as HIV-1 and can be fatal. Relatively speaking, however, it is much safer. The percentage of long-term nonprogressors (toward AIDS) is about 86%–95%, compared to less than 5% for persons with HIV-1. Generally, when reference is made to AIDS, this refers to HIV-1 [4].

The secret to curing HIV-1 lies in discovering its origin! It owes its genesis to the reality that multiple strains of SIVs came together inside a single cell and this allowed massive recombinant events to occur (that bypassed the miRNA defense system). In any human infected with HIVs (and there is always more than one strain or quasi-species of HIV that infects a person) the virus must bypass the unique miRNA defense system that is found in each of us. A successful infection is only possible if the initial HIVs are allowed to recombine *in situ* (inside each infected cells, as compared to *in vitro*, in a laboratory). If we can block the recombination of these initial HIVs, we can potentially block HIV from recombination and from developing thousands of quasi-species that one finds in a newly infected HIV individual. This may be one of the keys to prevent HIV-1. Little attention has been paid to this particular question during the past 35 years, while the pandemic has deepened, which makes discovering the origin of HIV all the more crucial now [12–15].

6.5 The search continues

Scientists still disagree on how AIDS was introduced into human populations in the twentieth century [4,5,16,17]. These things seem to us to be important to consider in continuing the search for the origin of AIDS. SIVcpzs, which infect nonhuman primates, are the ancestors of the human strain HIV type 1. This conclusion is founded on strong genetic sequence resemblance between the simian (SIVcpz) and human (HIV-1) strains that we have described above. On the other hand, HIV-2, a less pathogenic SIVsm lentivirus, traces its origin to sooty mangabeys (or white-collared monkeys) indigenous to Africa's western regions. SIVcpz is the main ancestral culprit that has provoked the particularly virulent HIV-1 strain that plagues humans. Of note, there is *no* physical evidence that HIV-1 existed before the OPV trial in Belgian Congo in the late 1950s, when a million or so people were inoculated with oral polio vaccine. Vaccines are usually helpful; perhaps this trial was an exception, a trial that inadvertently led to a global pandemic. We remind our readers that timing is important in determining causation. It does not appear to be a coincidence that the African continent has been particularly hard hit by HIV-1 [18].

References

1. Mats AN, Kuz'mina MN, Cheprasova EV. Viral contamination of polio vaccines in context of antivaccination mythology. *Zhurnal Mikrobiologii, Epidemiologii, i Immunobiologii* 2010;6:104–112.
2. Simon MA. Polyomaviruses of nonhuman primates: Implications for research. *Comparative Medicine* 2008;58(1):51–56.
3. Bagasra O. *HIV and Molecular Immunity: Prospect for AIDS Vaccine*. Natick, MA: Eaton Publishing, 1999.

4. Bagasra O, Pace DG. *Reassessing HIV Vaccine Design and Approaches: Towards a Paradigm Shift.* New York: Nova Scientific Publications, Inc., 2013.

5. Hooper E. *The River: A Journey to the Source of HIV and AIDS.* New York: Back Bay Books, 2000.

6. Zhu T, Korber BT, Nahmias AJ, Hooper E, Sharp PM, Ho DD. An African HIV-1 sequence from 1959 and implications for the origin of the epidemic. *Nature* 1998;391(6667):594–597.

7. Xie Y. Tracing the origins of HIV-1: Studying old samples of HIV strains she light on the history of the virus. http://arstechnica.com/science/2008/10/tracing-the-origin-of-hiv-1/, accessed August 6, 2016.

8. Addanki KC, Pace DG, Bagasra O. A practice for all seasons: Male circumcision and the prevention of HIV transmission. *Journal of Infection in Developing Countries (JIDC)* 2008;2:328–334.

9. Clutter DS, Jordan MR, Bertagnolio S, Shafer RW. HIV-1 drug resistance and resistance testing. *Infection, Genetics and Evolution* 2016. pii:S1567-1348(16)30369-0. doi:10.1016/j.meegid.2016.08.031.

10. Wang Y, Wu Y, Chen Y, Li C, Lu L, AuBuchon JP, Liu Z. The journey toward safer and optimized blood service in China: National strategy and progress. *Transfusion* 2016. doi:10.1111/trf.13773.

11. Beyrer C, Baral SD, Collins C, Richardson ET, Sullivan PS, Sanchez J, Trapence G, Katabira E, Kazatchkine M, Ryan O, Wirtz AL, Mayer KH. The global response to HIV in men who have sex with men. *Lancet* 2016;388(10040):198–206. doi:10.1016/S0140-6736(16)30781-4.

12. Putkonen P, Thorstensson R, Albert J, Hild K, Norrby E, Biberfeld P, Biberfeld G. Infection of cynomolgus monkey with HIV-2 protects against pathogenic consequences of a subsequent simian immunodeficiency virus infection. *AIDS* 1990;4:783–789.

13. Hakim ST, Alsayari M, McLean DC, Saleem S, Addanki KC, Aggarwal M, Mahalingam K, Bagasra O. A large number of the human microRNAs target lentiviruses, retroviruses, and endogenous retroviruses. *Biochemical and Biophysical Research Communications* 2008;369(2):357–362. doi:10.1016/j.bbrc.2008.02.025. Epub 2008 Feb 20.

14. Lowenstine LJ, Pedersen NC, Higgins J, Pallis KC, Uyeda A, Marx P, Lerche NW, Munn RJ, Gardner MB. Seroepidemiologic survey of captive Old-World primates for antibodies to human and simian retroviruses, and isolation of a lentivirus from sooty mangabeys (Cercocebus atys). *International Journal of Cancer* 1986;38(4):563–574.

15. NAM aidsmap. HIV-2. http://www.aidsmap.com/HIV-2/page/1322993/, accessed August 13, 2016.

16. Sharp PM, Hahn BH. The evolution of HIV-1 and the origin of AIDS. *Philosophical Transactions of the Royal Society of London B: Biological Sciences* 2010;365(1552):2487–2494. doi:10.1098/rstb.2010.0031.
17. Worobey M, Telfer P, Souquière S, Hunter M, Coleman CA, Metzger MJ, Reed P, et al. Island biogeography reveals the deep history of SIV. *Science* 2010;329(5998):1487.
18. Rutjens E, Balla-Jhagjhoorsingh S, Verschoor E, Bogers W, Koopman G, Heeney J. Lentivirus infections and mechanisms of disease resistance in chimpanzees. *Frontiers in Bioscience* 2003;8:d1134–d1145.

SECTION II
A Preventable Disease

7
AIDS
Choices and Outcomes

Ghajnuniet (AIDS in Maltese)

7.1 Like a worm in the vegetables

Acquired immune deficiency syndrome (AIDS) is ultimately a preventable disease, but its prevention must come from correct adult human choices, not from some magical scientific cancellation of consequences for bad behavior. The conduct that leads to human immunodeficiency virus (HIV) infection, and that indirectly afflicts innocent others who suffer through no bad choice of their own, is inconsistent with the major world religious traditions. In 1987, Ted Koppel, the famous and respected anchor of *ABC Nightline*, shared universal wisdom that has been much quoted since. It has become popular for the same reason that Tom Paine's *Common Sense* [1] appealed to American colonists at the time of the Revolution, or Harriet Beecher Stowe's *Uncle Tom's Cabin* [2] appealed to those who knew slavery was morally wrong and, through interaction with her book, solidified their own beliefs. Dr. Martin Luther King's famous "I have a dream" speech was all the more appealing because so many who did not share his eloquence did share his dream [3]. So it is with Koppel's statements at the 1987 Duke commencement: "We have actually convinced ourselves that slogans will save us. 'Shoot up if you must; but use a clean needle.' 'Enjoy sex whenever with whomever you wish; but wear a condom.'" He reminded the Duke audience that motives still matter: "No. The answer is no. Not no because it isn't cool or smart or because you might end up in jail or dying in an AIDS ward –but no, because it's wrong." He reminded his audience that "What Moses brought down from Mt. Sinai were not the Ten Suggestions, they are Commandments. Are, not were" [4].

Koppel's reminders are more than just Koppel's reminders; they echo teachings of major world religious traditions (including Islam, Judaism, Christianity, Buddhism, and Hinduism), which all revere the prophet Moses and the doctrines to which the news anchor referred. The eighth commandment recorded in the Chapter 20 of the Old Testament Book of Exodus [4] finds echo in the Chapter 17 of the Holy Qur'an [5], which characterizes adultery as both evil and shameful. The Jewish Talmud declares that "Immorality in the house is like a worm in the vegetables" (*Talmud, Sota 3b*) [5].

AIDS has become that worm in many houses in many lands. Hinduism also decries adulterous behavior, both on moral grounds and on grounds of caste confusion. Vishnu Purana 3:11 warns against immorality with another man's wife, teaches that even immoral thoughts are evil, states that adultery brings both mortal and postmortal punishment, and condemns the adulterous to rebirth later as an insect that creeps: "A man should not think incontinently of another's wife, much less address her to that end; for such a man will be reborn in a future life as a creeping insect. He who commits adultery is punished both here and hereafter; for his days in this world are cut short, and when dead he falls into hell" [5]. The Hindu Laws of Manu (Manusmriti) argues that "Men who commit adultery with the wives of others, the king shall cause to be marked by punishments which cause terror, and afterwards banish" [6].

Throughout history, positive behavior has been linked to adherence to community religious values. AIDS is not a disease that humans desire or plan to acquire; it is the result of irrational emotional behavior. Although the fear of suffering may deter HIV-inviting behaviors, compliance with the time-tested values of the world's great religious traditions provides an even more powerful deterrent. AIDS prevention efforts should not only focus on scientific logic but on the control of emotions. As HIV prevention is a matter of both heart and mind, both religion and science should be utilized as partners in prevention rather than mutually-exclusive antagonists. Although AIDS commonly traces its roots to unholy origins, the crusade against AIDS is one in which the peoples of the world must unite. A divisive holy war will not suffice; the world needs a unifying jihad against a killer that threatens people of all faiths, political persuasions, racial characteristics, and geographic origins. It is a particularly tough war to win because human volition is involved, and because most infected men or women do not know they are infected, and more than 80% of the underdeveloped nations do not have free HIV testing facilities. Nevertheless, when truly just and truly holy, wars can be worth fighting, and winning.

7.2 Historical perspective

Following the Spanish discovery of America by the Genoese Christopher Columbus, that nation, its Iberian neighbor Portugal, France across the Pyrenees Mountains, England nearby across the water, and other nations followed the great mariner's foot-in-the-door discovery with their own feet, legs, heads, and arms—those attached to the shoulders and those that European ammunition exploded on American soil. More potent and deadly than those menacing weapons, disease exploded on the New World scene. Although typically unintentional, various versions of *germ warfare* decimated native populations of the various Americas. This destruction-by-disease was an offshoot of the October 12, 1492 discovery, but it was not the fault of the Admiral of the Ocean Sea. The lack of understanding about how to prevent this disease was as profound, and more deadly, as the profound geographic ignorance that early explorers and settlers ignorantly shared. Frustration accompanied fear as disease mysteriously massacred native peoples. Lack of disease prevention has plagued every society; only the levels of sickness and mortality vary. It is difficult to war against a visible enemy and is terrible, but fighting invisible enemies is exasperating.

Fighting HIV has been both costly and lethal. But unlike the futile fight against smallpox, and other diseases, that post-Columbian Native Americans routinely lost, the modern-day battle against AIDS is one that could have been halted early, and then stopped dead in its morbid tracks. HIV, the viral foot-in-the-door that leads to subsequent colonization and exploitation by AIDS, has been seen under electron microscopes by scientists, and has been slowed by antiretroviral (ARV) cocktails (mixtures of drugs that fight the HIV retrovirus, and are known as HAART), but as of yet it has neither been prevented by vaccine nor cured by prescription [7].

The war against AIDS needs to be won, but it might best be thought of as a war that is won by avoiding the enemy. HIV is deadly once engaged. Why pitch to it, if routinely hits home runs? Why let it into *the paint* if it scores with slam-dunk regularity? Engaging in behaviors that tempt HIV has proven no more logical than pitching fastballs to Bryce Harper, or leaving LeBron James alone under the basket. Common sense still has its place. HIV is the mother of AIDS; human choice is the father of HIV. AIDS is, in reality, an illegitimate disease. If deprived of both father and mother, future illegitimate conceptions could simply be prevented.

7.3 Prevention: Male condoms

The likelihood of HIV infection through sexual transmission is vastly reduced when condoms are worn by men, as the Centers for Disease Control and Prevention (CDC) insists, *every time*. This does not totally eliminate the risk, but results in the laboratory and among humans overwhelmingly affirm the effectiveness of condoms in reducing sexually transmitted diseases

(STDs) including AIDS. The phrase *safe sex* is commonly used with regard to condoms but, in fact, they only contribute to *less risky* sex. The only true *safe sex* is either abstinence or having sex with only one partner in a relationship in which both partners are disease free and totally faithful to each other. The consequences of disloyal and unsafe sex, by men or women, are substantial and include the following: death of the individual participants; death to a baby born of an infected mother; or serious damage if the baby survives; severe damage to the brain, heart, and kidneys; sterility (loss of capacity to have children; and tubal pregnancies, which always results in the death of the fetus and sometimes that of the mother as well; cervical cancer for female partners; emotional difficulties; and difficulty in future intimate relationships.

Proper use of condoms by men not only reduces the risk of better known STDs (HIV and syphilis, gonorrhea, etc.), but other diseases, including Ebola and Zika virus, can also be transmitted sexually. Transmission takes place through bodily fluids (semen, blood, and vaginal fluid), and latex condoms have been found to be basically impermeable to even the tiniest of pathogens that cause STDs. Therefore, most of the emphasis regarding condoms has been placed on male-based solutions. Female-based solutions, covered in the next Section 7.4, are also receiving substantial attention and research funding, which is wisely given the reality that men often choose not to use condoms, even when they know that it is risky not to do so. Moreover, male condoms, while safe, are not foolproof.

Condoms have traditionally been used as a means of birth control, and they have been quite effective in accomplishing their purpose. This provides indirect evidence of the effectiveness of condoms in the prevention of sexually transmitted infections (STIs). Their effectiveness against STIs, including HIV, is still not fully understood. Research design to study the impact of condoms on STIs is fairly straightforward, but carrying out a rigorous research design is made very challenging due to practical and ethical considerations. To carry out the most meaningful experiments, humans are central to the process. This cannot be done without exposing uninfected individuals to potentially life-threatening danger or intentionally withholding treatment from infected individuals. Carrying out a thoroughly randomized experiment is virtually impossible under such circumstances. Instead, researchers must extrapolate information based on what they can find based on actual human behavior by previously infected and uninfected individuals whose behavioral patterns happen to become known to them. Surveys can help but knowing whom to survey, having the resources to carry out the research, and making statistically significant inferences are all daunting challenges [8–10].

7.4 Prevention: Female condoms
and drug-laced vaginal rings

Given the continuing challenges of blocking HIV infection in cases where males fail to take proper precautions not to infect their female spouse or partner, many pharmaceutical companies have turned to research about female condoms or other preventive measures that allow females to more directly care for their own health and that of potential offspring. One approach involves a protective vaginal ring that secretes ARV medication slowly and consistently. However, far from providing complete protection, the silicon band that is placed on the cervix has been found to lower HIV infection by only about 30%. Using the ARV drug dapivirine, the silicon-band rings are being developed to last longer. Additional sales potential is anticipated due to the combination of both anti-HIV and contraceptive properties in the same ring. In the past, vaginal creams with both antimicrobial and anti-HIV properties have been attempted, but the results have not been very favorable in terms of HIV prevention in women. Two major studies have been conducted, the first called ASPIRE and the second known as The Ring Study. Dapivirine was the experimental ARV drug used in these experiments carried out with 18–45-year-old women in the African nations of Malawi, South Africa, Uganda, and Zimbabwe. The efficacy rate was 27% in the first trial and 31% in the second. Although the argument can be made that some protection is better than none, the relatively low rates raise questions about why the rings are not more effective. Even more troubling were the results for the 18–21-year-olds in the study, they achieved a 0% efficacy level in the ASPIRE experiment and only a 15% level in The Ring Study. When these younger women were excluded, The Ring Study participants (ages 22–45) had a 37% efficacy level [11].

7.5 Who will protect the fetus from HIV?

Transmission of HIV from one partner to another is called horizontal transmission and that from a mother to her unborn child is vertical transmission. Progress in checking both the horizontal and vertical spread of HIV has been greatly enhanced by ARV medications. Vertical infection occurred in roughly 30% of cases in which a mother was HIV seropositive, which is particularly disappointing because mother-to-child transmission is now highly preventable. By far the leading manner in which children are infected by HIV is through perinatal transmission, from mother to child, during gestation, at the time of labor and delivery, or when the mother breastfeeds the new infant. HIV infection in infants has become an uncommon occurrence in the more developed parts of the world for several reasons. A higher percentage of women receive prenatal care, and governmental procedures commonly mandate an examination for HIV as part of this care. The CDC recommends early HIV testing for all pregnant women. In cases where infection is discovered, immediate prescription of an antiretroviral regimen imparts protection to the mother, keeping the virus in a dormant state, and either does the same thing with the fetus or outrightly guards the fetus from any HIV infection at all. The placenta plays a major role in blocking the HIV virus from entering the bloodstream of the fetus. HIV-positive mothers should provide alternative food sources to their babies and are encouraged to avoid breastfeed. However,if breast milk is the only safe food available to a child, as in many poorer nations with poor water purity, then breastfeeding is encouraged, coupled with appropriate ARV treatment. In HIV infected pregnant woman, C-section delivery may be chosen in order to minimize HIV transmission into the newborn's mouth, in cases where a child is at risk for HIV infection.

It is recommended that children infected by HIV begin ARV treatment during the first three months of life. This has been found to dramatically (about 75%) reduce deaths and the progression of the infection in these infants. Although HIV therapy is now universally recommended for children less than five years old, fewer than one of four even begins this life-saving and life-prolonging therapy due to lack of resources and medical care in poor nations. Troubling roadblocks in the treatment of these children have been HIV screening through organized diagnostic programs, and then, for those diagnosed as HIV-positive, and feasible arrangements for pediatric treatment on a long-term basis [12–14].

7.6 Sex with virologically suppressed HIV-infected individuals

Undoubtedly, HAART has contributed significantly to slowing HIV's spread across the earth. Still, the pandemic persists, particularly in poorer countries. The primary beneficiaries of ARV drugs, in terms of prevention, have been children, because HAART for expectant mother provides dual protection, for herself and for her fetus. This important protection is particularly important at the time of delivery, because that is when the newborn is at greatest risk of mother-to-child HIV transmission. If she has an elevated HIV viral level, or if her viral load is unknown prior to the time of anticipated birth, a cesarean delivery may be recommended to lessen the likelihood of vertical (mother-to-child) HIV transmission.

Given the high success rate in vertical transmission for HIV prevention, specialists realized that a similar protection could likely be afforded in horizontal cases (spouse-to-spouse, or partner-to-partner). Serodiscordant (or serodifferent) couples are those in which one individual is HIV-negative and the other HIV-positive. The Swiss National AIDS Commission took what at that time appeared to be a bold stand when they announced in 2008 that HIV cannot be transmitted sexually if the infected partner is taking ARV therapy, has a viral load that has remained below the detectable level (40 copies per milliliter) for at least six months, and has no other STIs. The announcement, known as the *Swiss statement*, was later clarified to explain that it did not mean there was no chance of HIV transmission but that the probability was so low that it was in the range of everyday risks, about 1 in 100,000. Although considered dangerous at the time, the Swiss statement was research-based and has proven to be sound. Very few HIV-transmissions have occurred among discordant couples. The analysis was based primarily on heterosexual unions, which still left the question whether men who have sex with men (MSM) would find the same level of protection [13–16].

A new study called the *PARTNER study* has sought to broaden the scope of serodiscordant pair analysis to include same-sex unions. Partners of People on ART (PARTNER) used a higher viral load threshold; 200 mL versus 40 mL in the largely heterosexual Swiss study, PARTNER has found that HAART, when correctly adhered to, virtually eliminates the likelihood of horizontal HIV transmission. A team led by Alison J. Rodger and Valentina Cambiano described a fundamental question regarding serodiscordant pairs: "A key factor in assessing the effectiveness and cost-effectiveness of antiretroviral therapy (ART) as a prevention strategy is the absolute risk of HIV transmission through condomless sex with suppressed HIV-1 RNA viral load (less than 200 copies/ml) for both anal and vaginal sex." In this study of 1,166 serodifferent couples that pulled together, evidence from 75 clinical sites located in 14 different European nations, the researchers found *no documented cases of within-couple HIV transmission* through sex without condoms. HAART, therefore, not only keeps HIV-infected individuals from developing AIDS, but also protects their partners. Additional security can still be achieved by using condoms [17,18].

7.7 Life-long anti-HIV drugs

HAART does not completely stop HIV replication. There are so-called *sanctuaries* that ARVs are unable to penetrate completely (e.g., the brain and the lymph nodes). In these sanctuaries, HIV can multiply and this proliferation can be measured by highly sensitive detection tools. Therefore, HIV infected individuals need to be committed to taking life-long ARV treatment. This is not a simple matter of taking HAART pills regularly and having the problem solved. Medicines generally have side effects, and ARVs have some serious ones. Patients whose viral load is very low may feel encouraged about blocking the progression of AIDS, but they may need additional medications for resolving the side effects of HAART. Moreover, opportunistic infections can accompany HAART regimens, patients can develop resistance to the HAART cocktails, and the antiviral combinations that have worked so well for a patient need to be changed periodically. In addition, there are certain HAART medications that cannot be used by pregnant women and young children [19].

It is really amazing that how ARVs have made a difference in the lives of HIV-infected individuals. In the early stages of the AIDS pandemic, there were no antiviral drugs and in the majority of the infected people, HIV infection basically presented a death warrant, as infected persons predictably died within a few years.

When should a patient start ARV therapy? Generally, HIV-1 seropositive individuals are those who have a CD4+ T-cell count in the range of 200–350 CD4+ T-cell per microliter of plasma. A healthy adult has a CD4+ T-cell count in the range of 500–1,600 cells per microliter (μL). Currently, controversy persists regarding when to commence a HAART regime. Some experts believe that a person living with HIV should be given HAART right away to avoid progression of the disease. This early initiation prevents viral secretion in the seminal fluid and vaginal secretions, and thus prevent further spread of the virus. A serious drawback is, however, that once ARV treatment is initiated it should never be stopped! HAART is not without consequences and side effects.

7.8 HIV medications approved by the Food and Drug Administration

So many individuals currently take or need ARV medication to keep their HIV infection from progressing to AIDS, for that we provide here a convenient summary for their reference, as well as for other concerned individuals. According to information summarized from the National Institute of Allergy and Infectious Diseases (NIAID), the National Library of Medicine (NLM), and the Food and Drug Administration (FDA), there are eight classes of drugs that fight HIV: (1) nucleoside reverse transcriptase inhibitors (NRTIs), (2) nonnucleoside reverse transcriptase inhibitors (NNRTIs), (3) protease inhibitors (PIs),

(4) fusion inhibitors, (5) entry inhibitors, (6) integrase inhibitors, (7) pharmacokinetic enhancers, and (8) a combination of HIV medicines. The first HIV drug class was the NRTIs, which was the only class available in the 1980s; Retrovir or AZT became available in 1987. In the 1990s, major strides forward occurred with the release of four different classes of HIV ARV medications: (1) NRTIs (four products), (2) NNRTIs (two new products), (3) PIs (four new products), and (4) a combination of HIV Medicines (one new product). In the first decade of the twenty-first century, seven of the eight HIV drug classes had products on the market, all except pharmacokinetic enhancers, which first made a product for sale in 2014 [20].

Given below is a brief description of each of the eight classes, with the brand names (with generic names and acronyms) under which they are marketed and sold, and given in order according to their year of approval by the FDA.

1. *Nucleoside reverse transcriptase inhibitors (NRTIs)*: In order to produce copies of itself, HIV relies on the enzyme called reverse transcriptase (RT). NRTIs block this enzyme and thus keep HIV from replicating. NRTIs are sold under the brand names *Retrovir* (zidovudine or AZT, 1987), *Videx* (didanosine, 1991), *Zerit* (stavudine, 1994), *Epivir* (lamivudine, 1995), *Ziagen* (abacavir, 1998), *Videx EC, enteric-coated* (didanosine, 2000), *Viread* (tenofovir disoproxil fumarate, 2001), and *Emtriva* (emtricitabine, 2003).

2. *Nonnucleoside reverse transcriptase inhibitors (NNRTIs)*: To replicate, HIV depends on the reverse transcriptase enzyme. NNRTIs bind to this enzyme and subsequently alter it in a way that keeps HIV from producing copies of itself, which helps prevent HIV from progressing to AIDS. NNRTIs are sold under the brand names *Viramune* (nevirapine, 1996), *Sustiva* (efavirenz, 1998), *Intelence* (etravirine, 2008), *Viramune XR, extended release* (nevirapine, 2011), and *Edurant* (rilpivirine, 2011).

3. *Protease inhibitors (PIs)*: Another enzyme HIV uses to assemble itself, after intracellular replication has taken place and all the proteins are translated for HIV to exit the host cells, is the protease. PIs are an effective anti-HIV drug because they block this enzyme. PIs are sold under the brand names *Invirase* (saquinavir, 1995), *Norvir* (ritonavir, 1996), *Crixivan* (indinavir, 1996), *Viracept* (nelfinavir, 1997), *Reyataz* (atazanavir, 2003), *Lexiva* (fosamprenavir, 2003), *Aptivus* (tipranavir, 2005), and *Prezista* (darunavir, 2006).

4. *Fusion inhibitors*: CD4 cells, important to the immune system, are protected by fusion inhibitors, which block HIV when it attempts to penetrate these CD4 cells. Fusion inhibitors are sold under the brand name *Zuzeon* (enfuvirtide, 2003).

5. *Entry inhibitors*: CD4 cells, important to the immune system, have proteins on them that HIV uses to enter these cells. Entry inhibitors, which block these proteins, are sold under the brand name *Selzentry* (maraviroc, 2007).

6. *Integrase inhibitors*: HIV uses the enzyme integrase to invade the host chromosomal DNA. Once HIV RNA is reversed transcribed by RT enzyme, its DNA migrates into the host nucleus. Integrase enzyme is essential for HIV DNA to enter the host genomic chromosomes. Integrase inhibitors, which can block this essential step of HIV replication, are sold under the brand names *Isentress* (raltegravir, 2007), *Tivicay* (dolutegravir, 2013), and *Vitekta* (elvitegravir, 2014).

7. *Pharmacokinetic enhancers*: HIV drugs that are already part of an HIV regimen are enhanced in their effectiveness by pharmacokinetic enhancers. These enhancers are sold under the brand name *Tyost* (cobicistat, 2014).

8. *Combination HIV medicines or HAART* (*highly active antiretroviral therapy*): Combinations of HIV medications have been found to be effective. These are defined as medicines comprised of at least two HIV medicines, which may come from a single drug class but generally are drawn from multiple classes, which blocks HIV at multiple steps (Figure 7.1). Combination HIV medicines are sold under the brand names *Combivir* (lamivudine and zidovudine, 1997), *Trizivir* (abacavir, lamivudine, and zidovudine, 2000), *Kaletra* (lopinavir and ritonavir, 2000), *Epzicom* (abacavir and lamivudine, 2004), *Truvada* (emtricitabine and tenofovir disoproxil fumarate, 2004), *Atripla* (efavirenz, emtricitabine, and tenofovir disoproxil fumarate, 2006), *Complera* (emtricitabine, rilpivirine, and tenofovir disoproxil fumarate, 2011), *Stribild* (elvitegravir, cobicistat, emtricitabine, and tenofovir disoproxil fumarate, 2012), *Triumeq* (abacavir, dolutegravir, and lamivudine, 2014), *Evotaz* (atazanavir and cobicistat, 2015), *Prezcobix* (darunavir and cobicistat, 2015), *Genvoya* (elvitegravir, cobicistat, emtricitabine, and tenofovir alafenamide fumarate, 2015), *Odefsey* (emtricitabine, rilpivirine, and tenofovir alafenamide, 2016), and *Descovy* (emtricitabine and tenofovir alafenamide, 2016) [20].

As just discussed, various combinations or cocktails of ARV drugs are currently available on the market. This may change as new combinations and the one-pill-a-day concept progress further. As we have seen, ARVs interrupt crucial steps in the HIV life cycle, including reverse transcription, integration of HIV cDNA in human chromosomal DNA, protease enzymes, and even attachment of HIV to CD4+ T-cell molecules and their coreceptors,

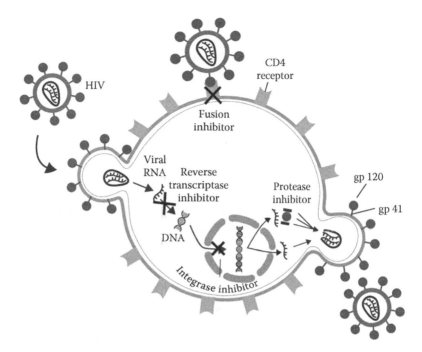

Figure 7.1   Molecular events in the HIV-1 life cycle and Interference by antiretrovirals. The HIV-1 gp120 envelope can bind to the CD4 receptors on the T-cell membrane. With the help of a coreceptor (CCR5 or CXCR4, not shown), the virion envelope fuses with the host membrane, and the virion core gains access to the cellular cytoplasm. However, fusion inhibitors can block this step and inhibit this viral entry. Once the virus enters inside the cytoplasm, one of the virion RNAs is reverse transcribed into cDNA and subsequently into dsDNA, which is known as preintegration complex. This two-step process can be blocked by two different classes of reverse transcriptase inhibitors. The HIV-preintegration complex can enter the nucleus and integrate into the host's genomic DNA. This step can be blocked by integrase inhibitors. Once integrated, the virus can begin to replicate by initiating transcription and translational machinery within the host and by producing a large quantity of the viral proteins and HIV RNA. The polyproteins produced by the viral genomes require proteases for final assembly. This step can be blocked by protease inhibitors. If a single type of inhibitor is used, the virus rapidly develops resistance. Therefore, multiple inhibitors are used to successfully block viral replication and reduce development of resistance.

called CXCR and CCR5. Certain combinations of HAART cocktails are so powerful that plasma HIV numbers (viral load) become so low that they are no longer detectable by conventional assays. Owing to the inhibition of HIV replication effected by HAART, the CD4+ T-cell count rises and the immune system remains intact, which results in a precipitous decrease in mortality and morbidity. In short, HAART had transformed HIV/AIDS into a chronic disease with which one may live for a long time.

7.9 Possible side effects from HAART treatment

HAART saves lives but it comes with some troubling side effects. Different classes of HIV drugs have varying side effects but the following, by way of summary, gives an indication of how diverse and troubling the spin-offs of HIV medications can be. Side effects include cardiovascular disease (including stroke), cholelithiasis (gallstones), diabetes and insulin resistance, dyslipidemia (including high cholesterol and triglyceride levels), gastrointestinal problems (including nausea, diarrhea, and vomiting), hypersensitivity (accompanied by headache, fever, fatigue, etc.), jaundice and hepatitis, kidney (renal) disease, kidney stones, lactic acidosis, lipodystrophy, loss of bone density, myopathy (muscle tissue disease), psychiatric and nervous system abnormalities, rash, and bleeding. As with other medications, health care practitioners and patients must balance benefits with risks. HAART lengthens out lives, but the quality of life is not the same as if the patient were not HIV-positive, and thus able to continue to live without medication [21].

7.10 Financial costs associated with HAART

According to recent estimates, the average monthly average wholesale cost for commonly prescribed ARV (HAART) medication comes is typically over U.S. $1,000 and runs as high as U.S. $4,098. The monthly cost for *NRTIs* ranged from U.S. $54–361 for Zidovudine (generic) to U.S. $1,197 for tenofovir disoproxil fumarate (Viread). The monthly cost for *NRTI combination products* ranged from U.S. $878 for Zidovudine/Lamivudine (generic) to U.S. $1,931 for Abacavir Sulfate/Zidovudine/Lamivudine (Trizivir). For the *Nonnucleoside reverse transcriptase inhibitors* (*NNRTIs*) class, the low was U.S. $648 for neviapine (generic) and the high U.S. $1,308 for etravirine (Intelence). For *Protease Inhibitors* (*PIs*), the low was U.S. $1,037 for lopinavir/ritonavir (Kaletra), and the high was U.S. $1,862 for darunavir/cobicistat (Prezcobix). For the drug class known as *integrase strand transfer inhibitors* (*INSTIs*), the lowest month cost was U.S. $1,445 for elvitegravir

(Vitekta) and the highest U.S. $3,415 for dolutegravir (Tivicay). For the *fusion inhibitor* category, the only cost listed was U.S $4,098 for enfuviritide (Fuzeon). For the *CCR5 antagonist* class, prices ranged from U.S $1,297 to U.S $2,594 for variations of the only product, maraviroc (Seizentry). For *Coformulated combination products as single tablet regimens*, the monthly low was U.S $2,815 for two products, rilpivirine/tenofovir, alafenamide/emtricitabine (Odefsey), and rilpivirine/tenofovir disoproxil fumarate/emtricitabine (Complera), and the monthly high was U.S. $3,245 for elvitegravir/cobicistat/tenofovir disoproxil fumarate/emtricitabine (Stribild). The least expensive rate for *pharmacokinetic enhancers (Boosters)* was U.S $231 for cobicistat (Tybost) and the most costly was U.S. $309 for Ritonavir (Norvir) [22].

References

1. Paine T. Common Sense. Great Books Online. http://www.bartleby.com/133/, accessed November 28, 2011.
2. Stowe HB. Uncle Tom's Cabin. The Literature Network. http://www.online-literature.com/stowe/uncletom/, accessed November 28, 2011.
3. King ML, Jr. I Have a Dream. American Rhetoric: Top 100 Speeches. http://www.americanrhetoric.com/speeches/mlkihaveadream.htm, accessed November 28, 2011.
4. Koppel, T. Koppel on Television and Morality. http://www.mediaresearch.org/mediawatch/1989/watch19890401.asp#analysis, accessed November 28, 2011.
5. World Scripture. Adultery. http://www.unification.net/ws/theme059.htm, accessed November 28, 2011.
6. Hindu Books Universe. Manusmriti: The Laws of Manu, 352. http://www.hindubooks.org/scriptures/manusmriti/ch8/ch8_351_360.htm, accessed November 28, 2011.
7. Bagasra O, Pace DG. *Reassessing HIV Vaccine Design and Approaches: Towards a Paradigm Shift*. New York: Nova Scientific Publications, Inc., 2013.
8. Weller S, Davis K. Condom effectiveness in reducing heterosexual transmission. *Cochrane Database Systematic Review* 2002;1:CD00325.
9. Centers for Disease Control and Prevention (CDC). Condom Effectiveness. http://www.cdc.gov/condomeffectiveness/, accessed August 12, 2016.
10. U.S. Food and Drug Administration (FDA). Condoms and Sexually Transmitted Diseases. http://www.fda.gov/ForPatients/Illness/HIVAIDS/ucm126372.htm#guar, accessed August 12, 2016.

11. Cohen J. Drug-laced vaginal ring succeeds against HIV – sometimes. *Science* 2016 Feb 22. http://www.sciencemag.org/news/2016/02/drug-laced-vaginal-ring-succeeds-against-hiv-sometimes, accessed August 15, 2016.

12. American Pregnancy Association. HIV/AIDS During Pregnancy. http://americanpregnancy.org/pregnancy-complications/hiv-aids-during-pregnancy/, accessed August 15, 2016.

13. Bhardwaj S, Carter B, Aarons GA, Chi BH. Implementation research for the prevention of mother-to-child HIV transmission in sub-Saharan Africa: Existing evidence, current gaps, and new opportunities. *Current HIV/AIDS Reports* 2015;12(2):246–255. doi:10.1007/s11904-015-0260-1.

14. Avert. Pregnancy, Childbirth and Breastfeeding and HIV. http://www.avert.org/hiv-transmission-prevention/pregnancy-childbirth-breastfeeding, accessed August 15, 2016.

15. AIDSinfo. HIV and Pregnancy: Preventing Mother-to-Child Transmission of HIV During Childbirth. https://aidsinfo.nih.gov/education-materials/fact-sheets/24/70/preventing-mother-to-child-transmission-of-hiv-during-childbirth, accessed August 29, 2016.

16. Avert. Fact Sheet: HIV & Mixed-Status Couples. https://www.avert.org/learn-share/hiv-fact-sheets/mixed-status-couples, accessed August 29, 2016.

17. Rodger AJ, Cambiano V, Bruun T, Vernazza P, Collins S, van Lunzen J, Corbelli GM, Estrada V, Geretti AM, Beloukas A, Asboe D. Sexual activity without condoms and risk of HIV transmission in serodifferent couples when the HIV-positive partner is using suppressive antiretroviral therapy. *JAMA* 2016;316(2):171–181.

18. NAM Aidsmap. No-one with an Undetectable Viral Load, Gay or Heterosexual, Transmits HIV in First two Years of PARTNER Study. http://www.aidsmap.com/No-one-with-an-undetectable-viral-load-gay-or-heterosexual-transmits-HIV-in-first-two-years-of-PARTNER-study/page/2832748, accessed August 29, 2016.

19. Chan CW, Cheng LS, Chan WK, Wong KH. Highly active antiretroviral therapy per se decreased mortality and morbidity of advanced human immunodeficiency virus disease in Hong Kong. *Chinese Medical Journal (Engl)* 2005;118:1338–1345.

20. For a convenient chart summarizing this information, see the source we have drawn on for our summary: AIDSinfo. HIV Treatment. FDA-Approved HIV Medicines. https://aidsinfo.nih.gov/education-materials/fact-sheets/print/21/0/1, accessed September 2, 2016.

21. AIDSinfo. Adverse Effects of Antiretroviral Agents. https://aidsinfo.nih.
    gov/guidelines/html/1/adult-and-adolescent-arv-guidelines/31/adverse-
    effects-of-arv, accessed September 2, 2016.
22. AIDSinfo. Guidelines for the Use of Antiretroviral Agents in HIV-1-
    Infected Adults and Adolescents. Clinical Guidelines Portal. https://
    aidsinfo.nih.gov/guidelines/html/1/adult-and-adolescent-arv-
    guidelines/459/cost-considerations-and-antiretroviral-therapy, accessed
    September 2, 2016.

8
Mosquitoes Do Not Spread HIV

VIGS (AIDS in Afrikaans)

8.1 Mosquito bites

HIV infection is integrally associated with the blood. Mosquitoes have taken a real interest in human blood. A mosquito bite occurs when a mosquito penetrates the human skin with what is called a proboscis, which is a type of miniature syringe. Female mosquitoes are the only mosquitoes that bite humans (male mosquitoes feel no biological need to do so; females need certain ingredients found in human blood to nourish their eggs) and sense a potential victim by an elaborate sensing system that can detect odor, carbon dioxide, light, and heat. Blood typically coagulates quickly, and thus checks an excessive human blood loss. Fortunately, from her viewpoint, mosquito saliva has anticoagulant proteins that keep human blood flowing by preventing clotting. Sucked blood enters the abdomen, and the undisturbed insect will continue ingesting the blood until her sensory nerve notifies her that the abdomen is full. Incidentally, a nonfunctioning sensory nerve would result in a blood-exploded mosquito [1].

8.2 Mosquitoes and human immunodeficiency virus

Could a female mosquito suck blood from an AIDS-infected person, and then spread HIV-1, the acquired immune deficiency syndrome (AIDS) virus, to its next blood provider? The short answer is *no*. The longer answer, also *no*, goes something like this: The amount of mosquito saliva is so small that it has no potential to infect another human. This does not mean that the concern for saliva-borne infection has been unfounded. Such historically lethal mega killers as malaria, yellow and dengue fevers, and now Zika virus, can be spread from person to person through mosquito saliva. Although humans have far more saliva than mosquitoes, and although human immunodeficiency virus (HIV) has been documented in saliva, it has only been detected in exceptionally low amounts. Human saliva, by itself, has never been the means of HIV transmission. Spitting may be disgusting, but it is not a means of transmitting HIV from the spitter to spitee, nor is it spread through kissing, when a large amount of saliva exchange occurs between mouths who happen to have open mouth sores or bleeding gums [2]. Interestingly, we were the first investigators to show the presence of HIV in oral mucosa [3]. Therefore, it is not that HIV does not infect the CD4+ cells in the mouth, but there are some components in saliva that prevent HIV from infecting the uninfected kissers.

8.3 Centers for Disease Control reports on insect transmission of HIV

In spite of extensive research, the influential institutional complex known as the CDC has reported zero-documented cases of AIDS transmission via a mosquito, or any other insect, for that matter [4,5]. This applies to areas that have plentiful supplies of both HIV-infected humans and mosquitoes. These biters, and other insects, do not share human blood or their own blood with subsequent bitees. When a female mosquito *bites* her victim, she does not share blood but rather injects a tiny quantity of lubricating saliva that enables her to feed more easily. HIV survives only briefly inside mosquitoes and cannot reproduce. There are no HIV-positive mosquitoes or insects. As deadly as they may be in passing along numerous other pathogens, they have no capacity to infect humans with HIV.

What about blood that remains on a mosquito's mouth parts; can it cause HIV infection? This is nothing to worry about. Mosquitoes rest and digest before looking for a new involuntary blood donor. By the time they extract more blood, they have no living HIV to share. Moreover, tiny mouth parts can host only tiny amounts of blood, and this too works against HIV-sharing by mosquitoes [2,4,5].

8.4 Mosquitoes and human mortality

Recent times have witnessed new threats to life, which have resulted in death by automobile accident, death by drug overdose, and premature death by overnourishment in a world plagued by malnutrition. There have been numerous deaths influenced by cars, bombs, and guns; and numerous premature deaths indirectly induced and cheered on by overnourishment, by excessive sugar intake, by mindless addiction to frivolous video games, and by lethargically immobile attendance (in front of ever larger video screens) at games that feature the vigorous physical exertions of talented, but at times steroid-loaded athletes. Yet it is critical to remember that cars, bombs, guns, food, sugar, and televised sports are not the ultimate cause of human suffering and death; humans are. It is they who drive the cars, drop the bombs, pull the triggers, feed themselves beyond reason, promote sugar cravings, and volunteer to be couch potatoes. Still, historically most deaths likely have not come through human choice, but through human incapacity to counter disease. The mosquito has been a leading actor on this stage of human life, and has unwittingly removed millions from the stage: (1) death by malaria, (2) death by yellow fever, (3) death by diarrhea, and (4) death of even very large persons by tiny mosquito proboscises.

The mosquito's needle-like *beak*, its blood-drawing instrument, intention-ally extracts blood and unintentionally infects with malaria and yellow fever, but not HIV. This tiny nemesis to human health seeks human blood for personal nourishment and for reproduction, yet has caused and continues to cause untold human suffering. Yellow fever, malaria, dengue, West Nile virus, and Zika virus are among the deadly examples of the power these miniature vampires have exerted over human life and history. Their effects are brutal yet their motives appear quite innocent: female mosquitoes extract blood to help them produce eggs (males have no such need and do not drink blood). The change of the exterior environment can menace mos-quitoes, stop their proliferation, and even stop their lives. For those who do become sick with malaria, it can be fatal but not always. Whether fatal or not, it will be miserable. The largely private world in which HIV spreads is not the world of mosquitoes.

HIV/AIDS is generally the result of human choices, although there are numerous innocent victims who had no choice in their own fate. For those who take the risks and who break the rules, it is not merely a matter of fate, but of selfishness, of unfaithfulness, of risk-taking, of exploitation, of disregard for societal norms, political laws, and moral values. It is behavioral fascism. Mosquitoes are active on other fronts, but they are not a direct threat on the AIDS killing fields. Humans are.

Mosquitoes, blood, and malaria; mosquitoes, blood, and yellow fever; mosquitoes, blood, and dengue. Why not mosquitoes, blood, and HIV? Mosquitoes can perform their blood-extraction routine just as easily with humans who are free from HIV as with those who are not. Mosquitoes do not discriminate in such matters. But whether a mosquito robs the blood of a totally uninfected person, one with a dormant HIV infection, or one with a full-blown case of AIDS, the HIV-transmission result is the same: *nada*. The mosquito cannot transmit HIV to humans. The quantity of infected blood ingested by the insect is simply too limited to allow for human-mosquito-human transfer of the HIV retrovirus. The very size of the mosquito, which facilitates its stealthy capacity to exact involuntary tribute-in-blood from unsuspecting victims, proves a decisive disadvantage in spreading AIDS. Troublemaking-via-mosquito, fortunately for humans, has its limits. Mosquitoes are, as it were, kept behind bars when it comes to spreading HIV.

In San José, Costa Rica, a tour guide commented on the proliferation of protective metal bars in front of house after house, and business after business: "We have a big problem with mosquitoes." Participants in the tour, including one of the authors of this book, laughed, and then heard the tour guide speak of two-legged mosquitoes that the metal bars are designed to deter. As useful as these unaesthetic bars may be in curtailing crime, more (not fewer) bars are needed to fight the unwelcome intruder known in Spanish as SIDA (síndrome de inmunodeficiencia adquirida; AIDS). Bars of societal norms, bars of political enactments, and bars of moral values must be reinforced, not weakened, by self-defeating approaches, if the two-legged mosquitoes that cause AIDS are to have their wings clipped.

References

1. Freudenrich CC. Mosquito Bites, Diseases and Protection. http://animals.howstuffworks.com/insects/mosquito3.htm, accessed May 26, 2011.
2. Centers for Disease Control and Prevention. HIV Transmission. http://www.cdc.gov/hiv/resources/qa/transmission.htm, accessed November 28, 2011.
3. Qureshi MN, Barr CE, Seshamma T, Reedy J, Pomerantz RJ, Bagasra O. Infection of oral mucosal cells by human immunodeficiency virus type I in seropositive persons. *Journal of Infect Diseases* 1995; 171:190–193.
4. Zuckerman AJ. AIDS and insects. *British Medical Journal (Clinical Research Ed)* 1986; 292(6528):1094–1095.
5. Booth W. AIDS and insects. *Science* 1987; 237(4813): 355–356.

9
Remedies That Are a Hoax

เอดส์ (AIDS in Thai)

What people believe influences how they act. In spite of overwhelming evidence that there is a global warming crisis, some continue to refuse to believe this conclusion, and they also refuse to examine the evidence. If a theater audience is persuaded by some illusion that an auditorium is on fire, the people in that audience will quickly exit. Desert travelers have unrealistically moved toward unreal mirages. Acts that accompany some false beliefs have inconsequential results but not so in the case of false notions regarding how to cure acquired immune deficiency syndrome (AIDS).

9.1 Ignoring HIV as the cause of AIDS

One particularly devastating belief about AIDS was pushed by politicians onto the general public of South Africa; they maintained that human deficiency virus (HIV) does not cause AIDS [1]. The consequences of this incredible position were catastrophic for this nation in which AIDS was firmly rooted and spreading quickly. University of California at Berkeley microbiologist Peter Duesberg popularized the notion that the HIV-AIDS link was phony [2–5]. Unfortunately, in spite of overwhelming evidence to the contrary, he has maintained his phony position, but not without severe consequences. Duesberg used his scientific stature, based on important cancer research in the 1970s, to argue that it was promiscuity mingled with drug abuse that caused AIDS. He has also argued that the antiretroviral (ARV) drugs prescribed to keep HIV-positive individuals alive combine with toxins in the environment to cause AIDS. Health officials in President Thabo Mbeki's South Africa who chose to adopt the temporarily less expensive position that served as a *scientifically*-supported excuse to avoid expenditures on ARV drugs. *Death by denial*, to use the journalist Brian Deer's phrase, was the result [6]. Even as South Africa's neighbors were funding antiretroviral therapy (ART) between 2000 and 2005, and saving lives, South Africa took a head-in-the-sand approach. Botswana transformed its AIDS care and its reputation between the 1980s, when it was regarded as the worst country in the world for AIDS, and the present, when about nine in ten HIV-positive individuals receives ART. By contrast, AVERT (AVERTing HIV and AIDS: AVERT goal is the provide information regarding HIV/AIDS, reduce stigma, provide information on transmission and treatment) estimates the rate of HIV-positive Americans currently taking ARV treatment as of 2011 at only 37%. The United States is the world's leading funder of anti-HIV medications but has serious problems in converting scientific knowledge into public awareness and enhanced awareness into improved compliance [6].

One of the authors of this book traveled to South Africa in 2006 as a part of HIV-prevention training. The country was in a very critical state indeed, with HIV denial resulting in numerous HIV-related funerals and burials. He also listened to Kary Mullis, a supporter of Duesberg whose invention of PCR earned him a Nobel Prize in chemistry, when he spoke at Thomas Jefferson University in 1992, Pennsylvania and asked him to explain the scientific bases of such an idea, given the fact that newborns were being born with HIV/AIDs and hemophiliacs were developing the disease after receiving tainted blood products [6]. There was no answer, just now the typical denial. Of note, Dr. Mullis has never worked with HIV virus [6,7]. Partly as a result of those dark days of denial, as of 2008, estimated third of a million South African adults have died of AIDS, and about 35,000 were already HIV-positive when they drew their first breath. The saying *ignorance is bliss* may be true in some instances, but in the case South Africa's official denial of the HIV-AIDS causal relationship, ignorance was simply ignorance, laced with costly denial that facilitated deferred mainte-nance to the body politic, for those bodies still alive.

9.2 False HIV cures and child abuse

In 2001, a Johannesburg journalist reported the shocking allegation that six South African men had raped an infant girl that was only nine-months old, that children had been the victims of a skyrocketing number of rapes and other sexual assaults, and that AIDS was a motivating factor. The utterly false notion that sexual activity with a virgin would cure a man with AIDS is yet another example of how the combination of desperation, depravity, and false beliefs can have catastrophic consequences. These cases shot from 37,500 in 1998 to 67,000 cases just one year later. As tragic as both num-bers are, child welfare advocates maintained that the real numbers could be as much as 10 times as high as the reported cases. The largest increase was in children between as young as six months and six years. Some infants died from injuries sustained in the sexual assaults. A decade and a half later, abuse of children and teenagers continues to be a problem, with 19.2% of the 15–17-year-old girls and 20.3% of the boys reporting having been a victim of some type of sexual abuse. Abuse does not cure disease; it spreads it and inflicts untold additional damage on its victims [8,9].

9.3 The electromagnetic pretense

Falsehoods have also taken their toll in other spheres, as in the marketing of the *Complete Cure Device*, promoted by the Egyptian military as a machine that would destroy HIV in the body through electromagnetism [10]. When a major general in the army boldly proclaimed that through divine grace he had a 100% cure for AIDS and one for hepatitis C as well, the scientific community within Egypt and beyond quickly saw the statement for what it really was an illogical scam. As announced, the bogus device would provide a cure within 16 hours by drawing a patient's blood, dismantling the disease and then readmitting the blood, now purified, into the patient. AIDS flows out, pure blood flows back in; easy enough to boast about, impossible to carry out. Still, the controversial proclamation was persuasive enough that 70,000 people wrote e-mails requesting access to the miracle machine. HIV is made from the same nucleotides (related to phosphate) and nucleosides (related to sugars), that are building blocks of all life forms. If HIV could somehow be blasted into harmless particles, then the whole human DNA and RNA would likewise be turned into microscopic rubble [10].

Another fake device sold in Johannesburg, South Africa was the so-called Angel Zapper. The small device is strapped to the ankle, an electric current is released, and the HIV is disabled by the 100 microampere electrical charges. This profoundly ridiculous idea is a hoax simply because the DNA or the RNA of the HIV virus is made of the same building blocks that make up all life forms and we, as humans, are consistently exposed to all sorts of electrical currents. Electrical currents are part of what make humans human. Externally administered electrical HIV-blasting charges may be financially beneficial to sellers, but they are worthless baggage when strapped to the ankles of financially-strapped HIV patients [11].

9.4 Spotted lizards, spotty research

The colorful Tokay Gecko, with its distinguishing orange spots, defining sounds, and bluish-grey skin has been promoted as an AIDS cure. With no scientific backing whatsoever, AIDS has been added to a list, which has grown over the years of maladies the tiny reptile is purported to cure, including cancer, diabetes, and asthma. Most HIV patients are poor, and wasting money is particularly tragic because ART medications are expensive. There is also an environmental concern. Demand has been so high (evidenced by the 1.2 million Tokay Gecko lizards exported annually from Java, more than 20 times the legal limit allowed for the pet trade) that environmentalist are concerned about the Tokay Gecko's long-time preservation [12].

9.5 Other quackery

The examples given are illustrative but certainly not intended to be comprehensive. Herbal remedies, special diets, and drinking alcohol are other purported cures for AIDS. Even high oxygen therapy, based on a regimen of bottled water mixed with hydrogen peroxide (H_2O_2), a toxic brew that could prove fatal. What researchers have found to be true has been reinforced by evidence presented throughout this book. HIV is caused, either directly or indirectly, by behavioral deviance. There is no cure; there is no vaccine. ART has progressed remarkably since the 1980s. For most HIV-positive individuals, AIDS can be a chronic disease that one can live with for many years [9].

References

1. Chigwedere P, Essex M. AIDS denialism and public health practice. *AIDS and Behavior* 2010;14(2):237–247.
2. Duesberg PH. HIV is not the cause of AIDS. *Science* 1988;241:514–517.
3. Duesberg PH. *Inventing the AIDS Virus*, pp. 169–216. Washington, DC: Regency Publishing Co., Inc., 1997.
4. Cartwright J. AIDS contrarian ignored warnings of scientific misconduct. *Nature*, May 4, 2010. http://www.nature.com/news/2010/100504/full/news.2010.210.html, accessed September 5, 2016.
5. Brink S. Fake cures for AIDS have a long and dreadful history. *NPR*. South Carolina Public Radio. August 17, 2014. http://www.npr.org/sections/goatsandsoda/2014/07/15/331677282/fake-cures-for-aids-have-a-long-and-dreadful-history, accessed September 5, 2016.
6. Johnson, G. Bright Scientists, Dim Notions. *The New York Times*, October 28, 2007. http://www.nytimes.com/2007/10/28/weekinreview/28johnson.html?_r=3&adxnnl=1&oref=slogin&ref=science&adxnnlx=1193583001-IE12EKQeJt1sjwCUOYPVWg&oref=slogin, accessed August 6, 2010.

7. Deer B. Death by denial: The campaigners who continue to deny HIV causes Aids. February 21, 2012. https://www.theguardian.com/science/blog/2012/feb/21/death-denial-hiv-aids, accessed September 5, 2016.

8. The Centre for Justice and Crime Prevention (CJCP) and the University of Cape Town's Department of Psychology and Gender, Health & Justice Research Unit (GHJRU). The Optimus Study on Child Abuse, Violence and Neglect in SA. July 31, 2015. https://www.uct.ac.za/usr/press/2015/OptimusStudy_31July2015.pdf, accessed September 5, 2016.

9. Flanagan J. South African men rape babies as "cure" for aids. *The Telegraph*, November 11, 2001. http://www.telegraph.co.uk/news/worldnews/africaandindianocean/southafrica/1362134/South-African-men-rape-babies-as-cure-for-Aids.html, accessed September 5, 2016.

10. Abdelaziz S, Abedine S. Egyptian army's AIDS-cure claim gets harsh criticism. *CNN*, February 27, 2014. http://www.cnn.com/2014/02/27/world/africa/egypt-aids-cure-claim/, accessed September 5, 2016.

11. Bhardwaj V. A roundup of fake AIDS 'cures': Angel Zapper, Garani MW1, Topvein, SF 2000. *Africa Check*, July 20, 2016. https://africacheck.org/reports/a-roundup-of-fake-aids-cures-angel-zapper-garani-mw1-topvein-sf-2000/, accessed September 5, 2016.

12. Unfounded Claims of HIV cure spike demand for Tokay Gecko in Southeast Asia. November 16, 2011. http://www.asianscientist.com/2011/11/health/tokay-gecko-southeast-asia-hiv-cancer-cure-2011/, accessed September 5, 2016.

10
Circumcision

10.1 ABC or ABCC?

Countries and organizations worldwide have stressed three easy-to-remember basic prevention practices, the ABCs of controlling the deadly human immune deficiency virus (HIV) retrovirus: abstinence, be faithful, and condoms [1]. Although the ABC approach seems logical and simple, and may have been helpful in saving many lives, it has not proven to be either universally logical to researchers or appropriately simple to learners. In 2006, the *Washington Post* reported that the approach knows ABC as an "AIDS-prevention strategy widely promulgated both here and abroad, got a distinctly mixed report card as African countries reported their experiences to delegates at the 16th International AIDS Conference (in Toronto)" [2]. A study of black youth of junior high school age found that these young people were actually less likely to instigate sexual relations if taught only premarital abstinence than if given a fuller A-through-C curriculum on acquired immunodeficiency syndrome (AIDS) prevention. A study of safe anti-HIV practices in Botswana found a disturbing disconnect between knowledge and practice. CDC-sponsored efforts to educate people seem to have succeeded in raising educational awareness but not in changing actual behavior. The director of the joint US–Botswana collaboration, dubbed *BOTUSA*, reported that "People who were exposed to the program had greater knowledge but were no more likely to be practicing ABCs." The ABC approach, on the other hand, seems to have been much more successful in Uganda [2].

In contrast to a mere hand-wringing recitation of statistical HIV woes, we suggest that the helpful ABC guidelines should be carefully adapted to the specific groups being served. Perhaps a group of middle school Hispanics needs less be-faithful-to-your-spouse encouragement, and a decreased emphasis on condom use, than instruction on the moral and practical values of postponing sexual relations until marriage. Although many will already have had such intimate experiences, and even more will likely to have them as they become older teens, they may well have fewer such experiences, and be more careful in partner selection than if they received no instruction. Some will even practice full abstinence. The dilemma of under-versus-over instruction must be faced wisely by local teachers and administrators. Too little instruction could prove deadly, whereas excessively explicit instruction may actually inflame already highly inflammable emotions.

ABC, then, should be used judiciously. We propose that a second C, for Circumcision, should also be considered by educators, health care providers, and policy makers alike. The titles of some of the now substantial literature on the subject give some idea of the global interest in the confluence of circumcision and HIV prevention: (1) "The relationship between male circumcision and HIV infection in African populations" [3], (2) "Male circumcision: an acceptable strategy for HIV prevention in Botswana" [4], (3) "Neonatal Circumcision: A Review of the World's Oldest and Most Controversial Operation" [5], (4) "Randomized, controlled intervention trial of male circumcision for reduction of HIV infection risk: The ANRS 1265 trial" [6], (5) "HIV-1 target cells in foreskins of African men with varying histories of sexually transmitted infections" [7], (6) "Male circumcision for HIV prevention in young men in Kisumu, Kenya: a randomized controlled trial" [8], (7) "Male circumcision, religion, and infectious diseases: an ecologic analysis of 118 developing countries" [9], (8) "The potential impact of male circumcision on HIV in sub-Saharan Africa" [10], and (9) finally, an article to which we contributed in 2008, "A Practice for All Seasons: Male Circumcision and the Prevention of HIV Transmission" [11].

Awareness of the global picture can help us attack the problem at local and national levels. On the global stage, HIV/AIDS has reared its ugly head repeatedly. Its Medusa-like influence poses herculean challenges that seem unstoppable. It shows no favoritism on grounds of race, economics, social status, or gender. We must always remember that behavioral patterns lie at the root of the epidemic, and yet we must also recognize that there are striking differences in prevalence rates among those with similar patterns of high-risk behaviors. Male circumcision clearly exerts a positive influence in the reduction of HIV infection. As we (Professors Addanki, Pace, and Bagasra [11]) have written elsewhere, by way of summary, male circumcision (MC) "is known to significantly reduce female-to-male HIV transmission through sex, which then decreases male-to-female transmission. Three recent randomized controlled studies from Africa have shown that circumcision offers a 60%–70% protective effect against heterosexual acquisition of HIV" [11]. We explained that "(the protective effect of circumcision against HIV, known since the 1980s, has been confirmed by more than 30 studies before…three famous randomized controlled trials, which are the criterion standard of clinical research.)" We reported that there is a dramatic reduction in HIV prevalence in those countries with a rate of circumcision that exceeds 80% [3,10]. Moreover, we found what others have also discovered: "MC not only reduces HIV but also other sexually transmitted diseases (STDs)." The convincing nature of the evidence has led to endorsement of male circumcision by leading world health institutions, including the World Health Organization (WHO) and the National Institutes of Health (NIH) [11].

10.2 Early childhood circumcision versus adult circumcision

According to Brian J. Morris, newborn circumcision has been on the increase in the United States. The typically high rate for Whites has remained high, and the rate for Blacks has increased: "The rates recorded in the north-east region were steady at 70%, while rates rose in the mid-west (80%) and South (70%). For the western region rates have been falling due to the influx of Hispanics (50% of all births, so diluting out the overall rate in California to 35%)." He reported that "overall the statistics show an increase in circumcision rates for Non-Hispanic Whites," and explained that "in the West individual hospital data have shown…the rate for non-Hispanic Whites is in fact 75%–80%." An alarming finding, however, is that "for the next generation of Hispanics, (only) 29% of boys are circumcised" [12].

10.3 Educating community regarding the benefits of circumcision

The answers to the troubling rates of infection for Hispanics lie, at least partially, in educational and religious instruction, in both private and public settings. As immigrant Hispanics are more fully integrated into America's educational institutions, one would expect higher rates of positive health practices, including circumcision. Muslim and Jewish circumcision rates are very high, for religious reasons, but the persuasion of other groups may need to focus more on matters of health. Education, both in the schools and in the media, plays a major role in this arena. Where cost plays a role in the decisions about the circumcision of newborns, public policy makers might consider whether it would be in the best economic interests of the public to subsidize circumcision expenses, or pay for them entirely. They may also want to consider whether free circumcision for males of any age would be wise, on both health and economic grounds. Numerous Latino immigrants have migrated illegally, a problem compounded by their poverty. To pay an extra U.S. $1,000 to circumcise a male newborn is prohibitively expensive for them. Free state-supported male circumcision makes more sense, and would be money well spent by state and national governments.

There are ways to tailor educational efforts to specific groups. Surely, our advertisers and scholars could suggest promising methodologies. As country of origin seems to make a difference, why not give additional special attention to the most at-risk groups? If language barriers increase the probability of becoming HIV seropositive, why not make English language education more accessible (for a variety of reasons), and promote more bilingual communication. These are daunting challenges, and yet improvements can surely be made. Other challenges involve the role that hegemonic male–female power relationships, including *machismo* [13] have on seropositivity, and how disparities in health care relate to HIV to AIDS progression. The health implications of double-standard *machismo* infidelity among Mexican men is serious, as Hirsch et al. conclude in their analysis of HIV factors for rural women in Mexico: "Marriage presents the single greatest risk for HIV infection among women in rural Mexico" [14]. As drug and alcohol abuse contribute to the health challenges Hispanics face, would not stepped-up antialcohol educational efforts be warranted? Are not stricter controls on advertising alcohol a means of fighting battles against cheaper HIV roots rather than wars against costly AIDS branches? Ultimately, AIDS is curable only to the degree that human behavior is improved. Given the bleakness that darkens the contemporary Hispanic health horizon, the partial solution emphasized in this paper, the MC solution, seems all the more inviting. Male circumcision may, in fact, be a realistic *behavioral vaccine* that can no longer be overlooked.

Beneficial health practices often coincide with religiously motivated behaviors. The use of tobacco was prohibited or discouraged by some religions long before United States surgeons-general began to routinely denounce tobacco as both addictive and deadly. Similarly, religious opposition to alcohol use predated the persuasive scientific data about the dangers of this substance, which is now commonly recognized as a drug. Male circumcision, now commonly performed for health reasons, also has deep religious roots. The removal of the male foreskin is an ancient practice that is well-documented in the Christian Bible (*circumcision*) [14], the Jewish Torah (Pentateuch) (*bris*) [15], and the Muslim Sunnah (*tahara*) [16]. This practice is by no means universal, and is not practiced by some major religious groups including Hindus, Buddhists, and Sikhs [17].

10.4 Irrational opposition to circumcision

Owing to historical rivalry and even conflict between some religious groups, the decision to avoid circumcision can be both a matter of adherence to the practices of their own group and the rejection of the practice of the rival group. Hindu–Muslim tension provides a case in point. Islam continues to preach circumcision as an inspired practice that goes back to the patriarch Abraham. Hinduism has no such doctrinal or behavioral link to either Abraham or circumcision. Yet for a Hindu to reject circumcision because it is a *Muslim practice* would perhaps be understandable, but not sustainable on logical grounds. Numerous physicians and others perform male circumcision across the world, and these medical procedures may or may not have anything to do with religious doctrine or tradition.

The debate over circumcision might be but one of many instances of how religions differ in their doctrines and traditions. The debate has moved beyond a mere matter of religion, however, as confirmed by the growing body of scientific evidence that links circumcision to lower rates of HIV and AIDS. Circumcision is good for the human body, and what is good for the human body, whether considered good or not by a particular religious group, is good for the body politic.

Circumcision has proven to be a remarkably effective deterrent to infection, so much so that it may be characterized as a vaccine by minor surgery. Even without considering sanitation benefits, penile and cervical cancer prevention [18, pp. 50–61], and other health benefits, male circumcision is so successful in HIV prevention as to merit consideration by public policy makers at all governmental levels. It is important to note that once a circumcised male has become infected with HIV, he is as likely to transmit the virus to someone else as is an uncircumcised male. Females are protected by male circumcision; therefore, by the lower incidence of infection that circumcision affords [8, pp. 178–180]. Medical procedures cost money, and poverty can and does present an obstacle to proper care; so does ignorance. These two obstacles can be challenging hurdles for persons of any economic class, but for the poor they may be high hurdles indeed. Public funding may well be needed to lower or remove all barriers to circumcision.

To pay for an adult circumcision may seem uninviting to a male who is unconvinced about the utility of circumcision for himself. Funding agencies, public and private, may do well to consider providing financial incentives to male adults who agree to be circumcised. At birth, circumcision is generally easiest and cheapest. It also makes monitoring most feasible.

An HIV-related *tale of two cities* makes a strong case for intervention via circumcision, especially early intervention.

Lozi and Luvale are African villages located in the same valley in southern African nation of Zambia. Less than a mile separates them from each other, and their lifestyle patterns and sizes were likewise close. A major difference, one with profound health implications, separated them: the men of Luvale were circumcised for religious reasons; the men of Lozi were not. Early in the 1980s, the HIV/AIDS epidemic threatened the area. Villagers soon noticed that HIV infection rates were far higher in Lozi, the town with the uncircumcised men. One-fifth of the young adults in the town tested positive, whereas their counterparts in Luvale the rate was only about one-third that rate (7%). A calculated projection was made that if something were not done, Lozi would lose a staggering three-fifths of its children to AIDS. In spite of local tradition and the anticircumcision position, village elders and parents from Lozi commenced taking their children elsewhere to have them circumcised [18, p. 36].

This pattern symbolizes what is taking place across the world: circumcised men, like the men of Luvale, are far safer when it comes to HIV/AIDS. What is true locally, as seen in the previous example, is also true globally. A comparison of nations where circumcision is common with those where it is not shows a remarkable difference in HIV infection rates. Circumcision is a matter of individual choice, and such choices are more likely when adults are educated about the benefits they and their families and their communities can receive by choosing circumcision. The circumcision decision is best made by someone other than the recipient of the procedure; infants cannot make this choice but it should be made for them, just as the decision to provide proper nutrition is made for them. On account of the vast numbers of uncircumcised males, the ideal public-policy approach would be to encourage, with governmental financial backing as needed, adult male circumcision, and infant circumcision as well. Were this to take place, even without improvement in public morality, HIV infection rates would plummet.

Considering the enormous monetary sums that have been spent searching for a still elusive AIDS vaccine, public and private planners would do well to consider providing major funding to educate and persuade the populations of the world to promote circumcision.

Thus far, we have focused on the advantages of circumcision for males, but the benefits to females are also substantial. HIV-free males do much to keep wives and illegitimate partners free of infection as well. Cervical cancer rates are far lower for women when the men with whom they are intimate are circumcised. When performed with proper medical procedures, circumcision basically has a positive benefit with no negative side effects. It is, as we have argued elsewhere, a *practice for all seasons* [11]. At the levels of both home and planet, male circumcision would improve both male and female health across the earth. Now is the time to plan globally, and circumcise locally.

References

1. AVERT. The ABC of HIV Prevention. http://www.avert.org/abc-hiv.htm, accessed November 29, 2011.

2. Brown D. Africa Gives 'ABC' Mixed Grades: AIDS Abstinence Plan Raises Awareness but Has Small Effect on Behavior. *Washington Post*, A04. August 15, 2006. http://www.washingtonpost.com/wp-dyn/content/article/2006/08/14/AR2006081401458.html, accessed January 20, 2009.

3. Bongaarts J, Reining P, Way P, Conant F. The relationship between male circumcision and HIV infection in African populations. *AIDS* 1989;3:373–377.

4. Kebaabetswe P, Lockman S, Mogwe S, Mandevu R, Thior I, Essex M, Shapiro RL. Male circumcision: An acceptable strategy for HIV prevention in Botswana. *Sexually Transmitted Infections* 2003;79:214–219.

5. Alanis MC, Lucidi RS. Neonatal circumcision: A review of the world's oldest and most controversial operation. *Obstetrical & Gynecological Survey* 2004;59(5):379–395.

6. Auvert B, Taljaard D, Lagarde E, Sobngwi-Tambekou J, Sitta R, Puren A. Randomized, controlled intervention trial of male circumcision for reduction of HIV infection risk: The ANRS 1265 trial. *PLoS Medicine* 2005;2(11):e298. Epub 2005 Oct 25. Erratum in: PLoS Med. 2006 May;3(5):e298.

7. Donoval BA, Landay AL, Moses S, Agot K, Ndinya-Achola JO, Nyagaya EA, MacLean I, Bailey RC. HIV-1 target cells in foreskins of African men with varying histories of sexually transmitted infections. *American Journal of Clinical Pathology* 2006;125:386–391.

8. Bailey RC, Moses S, Parker CB, Agot K, Maclean I, Krieger JN, Williams CF, Campbell RT, Ndinya-Achola JO. Male circumcision for HIV prevention in young men in Kisumu, Kenya: A randomized controlled trial. *Lancet* 2007;369:643–656.

9. Drain P, Halperin D, Hughes J, Klausner J, Bailey R. Male circumcision, religion, and infectious diseases: An ecologic analysis of 118 developing countries. *BMC Infectious Diseases* 2006;6:172.

10. Williams BG, Lloyd-Smith JO, Gouws E, Hankins C, Getz WM, Hargrove J, De Zoysa I, Dye C, Auvert B. The potential impact of male circumcision on HIV in sub-Saharan Africa. *PLoS Medicine* 2006;3:e262.

11. Addanki KC, Pace DG, Bagasra O. A practice for all seasons: Male circumcision and the prevention of HIV transmission. *Journal of Infection in Developing Countries* 2008;2:328–334.

12. Morris BJ. Male Circumcision Guide for Doctors, Parents, Adults & Teens. http://www.circinfo.net/rates_of_circumcision.html, accessed February 5, 2009.

13. Reproductive Health Issues in Latin America. http://www.stanford.edu/group/womenscourage/Repro_Latin/ekobash_Homepage_Latin.html, accessed November 30, 2011.

14. Hirsch J, Meneses S, Thompson B, Negroni M, Pelcastre B, Rio C. The inevitability of infidelity: Sexual reputation, social geographies, and marital HIV risk in rural Mexico. Framing health matters. *American Journal of Public Health* 2007;97(6):986–996. http://ajph.aphapublications.org/doi/full/10.2105/AJPH.2006.088492, accessed November 30, 2011.

15. The Hebrew Bible in English. http://www.mechon-mamre.org/e/et/et0.htm, accessed November 30, 2011.

16. Rizvi SAH, Naqvi SAA, Hussain M, Hasan AS. Religious circumcision: A Muslim view. *BJU International* 1999;83(Suppl. 1):13–16.

17. BBC Religions. Religions. Circumcision of boys. http://www.bbc.co.uk/religion/religions/islam/islamethics/malecircumcision.shtml, accessed November 30, 2011.

18. Schoen E. *Circumcision*. Berkeley, CA: RDR Books, 2005.

11
Clean Needles
The Case of India

СИДА (AIDS in Serbian)

11.1 Sterilization practices

Clean needles are troublingly unavailable in India. Particularly disturbing is the pattern by which low-level staff members hoard clean needles, only to sell them on the black market. Even HIV/AIDS clinicians use reusable needles that are *sterilized* in warm water. This appears to be a widespread practice throughout India. In addition, tattooing and shaving by barbers with unclean blades are common practices in India that can contribute to human immunodeficiency virus (HIV) transmission. We maintain that the widespread use of relatively contaminated needles for basic medical procedures is one of the major culprits in the spread of HIV. These include the use of such unsafe needles even by the staff members of physicians and nurses, who generally are uneducated in sterilization techniques [1]. We propose that Indians should be educated in this urgent matter by mass media advertisements. As a significant percentage of Indians are uneducated, they need to receive information by other means than reading brochures or handouts (preferably by TV). This will require actual visual images from TV and movie previews. In addition, people need to be educated about other potential means of getting infected than through sexual activity, including tattooing, body piercing, and intravenous drug use.

11.2 Tatooing or body piercing

India's National AIDS Control Organization (NACO) places tattoos among its HIV risk factors. It warns against *Injections, tattoos, ear piercing, or body piercing using nonsterile instruments* [2]. The Centers for Disease Control (CDC) in the United States also warns about the potential of people becoming acquired immune deficiency syndrome (AIDS) infected through tattooing. There is a risk that the HIV could be transmitted *if instruments contaminated with blood are either not sterilized or disinfected nor are used inappropriately between clients.* CDC advises single use only, followed by safe disposal, of all *single-use instruments intended to penetrate the skin.* It warns that "reusable instruments or devices that penetrate the skin and/or contact a client's blood should be thoroughly cleaned and sterilized between clients" [4]. Those who *do tattooing or body piercing should be educated about how HIV is transmitted* and follow procedures that will *prevent transmission of HIV and other blood-borne infections in their settings.* According to the CDC, before getting a tattoo or subjecting their body to any piercing, individuals should directly "ask staff at the establishment what procedures they use to prevent the spread of HIV and other blood-borne infections, such as the hepatitis B virus" [3,4].

Obviously, the safest and cheapest course to take is to simply avoid all tattooing and body piercing. However, this hardly seems practical in contemporary India because virtually all Indian girls get their ears and right side of their nose pierced. Hundreds of millions of people are involved; sterilization of needles, therefore, is imperative.

11.3 Need for institutional involvement

The major efforts of Indian governmental authorities are directed toward a few states that appear to have the highest incidence of HIV infections. The alarm regarding these areas is understandable, but given the finite resources available to fight HIV/AIDS, care must be taken to balance prevention and treatment, and also to the balance the relative amounts of aid going to the various regions of India. HIV prevalence data for each state are primarily established through antenatal clinics, where pregnant women are tested. Although these data only directly reveal statistical information for sexually active women, they can have general interpretive value when intelligent data interpolation takes place. Much more needs to be done to *prevent* rather than *react* to existing problem areas. Greater advocacy by higher education (including students, who themselves comprise a high-risk group) and by NGOs (who need greater accountability [5], motivation, and training materials) is needed. Although these groups are making some genuine efforts, the HIV *crisis* requires an increased sense of urgency. Higher education has a unique, but unfulfilled, opportunity to play a leading role in the fight against HIV/AIDS. Administrators, faculty members, and students hold an important key in the effective dissemination of accurate information to their local communities. With their linguistic expertise, social status, political clout, and communication potential, the higher education community has unparalleled potential to teach India what it must do to avert even more serious disaster.

Given the complexities of Indian culture, a multipronged response is needed to attack the multifaceted crisis. Such a response must urge an attack on AIDS that is conducted in a culturally sensitive manner. Given the variety of cultures in the country, we might more accurately speak of *Indias* and of HIV/AIDS *crises*. Each of these *Indias* would benefit from a systematic, nationwide approach to prevention and cure, but in the national and global war against HIV/AIDS, each battle must ultimately be fought on a separate and more localized battleground by forces that are linguistically and culturally capable of confronting the epidemic in ways that are culturally-sensitive.

References

1. Pace DG, Bagasra O. *Reflections on South Asia: Private Wants or Community Needs?* Saarbrücken, Germany: VDM Verlag Dr. Müller Aktiengesellschaft, 2010.
2. National AIDS Control Organisation. Antiretroviral Therapy Guidelines for HIV-Infected Adults and Adolescents Including Post-exposure Prophylaxis. Ministry of Health and Family Welfare, Government of India 2007, with support from CDC, Clinton Foundation, WHO. http://www.ilo.org/wcmsp5/groups/public/---ed_protect/---protrav/---ilo_aids/documents/legaldocument/wcms_117317.pdf, accessed December 13, 2011.
3. Flores GL, de Almeida AJ, Miguel JC, Cruz HM, Portilho MM, Scalioni Lde P, Marques VA, Lewis-Ximenez LL, Lampe E, Villar LM. A cross section study to determine the prevalence of antibodies against HIV infection among Hepatitis B and C infected individuals. *International Journal of Environmental Research and Public Health* 2016 Mar 11;13(3). pii: E314. doi:10.3390/ijerph13030314.
4. Centers for Disease Control and Prevention. HIV Transmission. http://www.cdc.gov/hiv/resources/qa/transmission.htm, accessed November 28, 2011.
5. Pace DG, Bagasra O. NACO and the World Bank are correct in their crackdowns. *Nature Medicine* 2008;14:588. doi:10.1038/nm0608-588.

SECTION III
What to Do and Not Do

12
Alcohol and AIDS

SCDI (AIDS in Latin)

12.1 Alcohol's threat to world health

The World Health Organization (WHO) has warned of an orally ingested substance that plagues global health, and causes more deaths annually than what acquired immune deficiency syndrome (AIDS), tuberculosis, or violence does. That substance is alcohol, and it causes almost 4% of all deaths on the planet. Male alcohol-related deaths are particularly pronounced: "Globally, 6.2% of all male deaths are attributable to alcohol, compared to 1.1% of female deaths" [1]. The percentage for males in the Russian Federation and countries near its borders is much higher, approximately one out of every five. Globally, for about one in 11 individuals of those who die between the ages of 15 and 29 years, the cause of death is related to the use of alcohol [1]. Although alcohol use caused more deaths than any of the other three global killers, it is not possible, as the WHO recognizes to artificially isolate alcohol: "Harmful drinking is also a major avoidable risk factor for noncommunicable diseases, in particular cardiovascular diseases, cirrhosis of the liver, (inflammation of the pancreas) and various cancers. It is also associated with various infectious diseases like HIV/AIDS, (STDs) and TB, as well as road traffic accidents, violence, (homicides) and suicides" [2]. Those who choose alcohol typically choose much more than a meager drink. They also often invite terrible personal distress, health problems, family grief, physical suffering, emotional pain, and premature death. As terrible as is the toll that alcohol inflicts on the health of those who choose to drink it, the drug alcohol also inflicts terrifying consequences on mind and spirit [1].

12.2 Prosperity, alcoholism, and AIDS

Human choice has so much to do with human behavior, and the choice to consume alcohol is notorious for impairing future choices. Indeed, at times, it is the choice to end all choices. The propensity to increase alcoholic intake has been linked to higher incomes in India, South Africa, and other African and Asian countries with sizeable populations [3]. It is important to remember, however, that although a pay rise may make purchases more possible, it is not the money that makes the decisions; humans do. It is the increasing frequency of destructive decisions vis-à-vis alcohol that causes such toxic pairings as alcohol-AIDS, alcohol-tuberculosis, and alcohol-accidents. Still, public-alcohol policies commonly have not been elevated to a high-priority position on public-policy agendas, even as elevated blood alcohol, accident, TB, and human immunodeficiency virus (HIV) levels increasingly clamor for their attention. And this is only the short list; the longer enumeration includes child neglect, poor job performance, lost employment, spouse abuse, unhappy relationships, violence, and the types of immoral behaviors that invite HIV into bloodstreams, homes, and communities. Public officials have the duty to balance individual freedom with the public welfare, and it is in no one's best interests to table serious policy discussions about alcohol/AIDS.

Scholars have long realized that a substantial number of HIV-infected persons drink alcohol. Although they continue to narrow in on the precise impact of alcohol on HIV proliferation, the consensus is that the news is bad. One study of Indian rhesus macaques found that after 18–24 weeks following HIV infection, the group with alcohol had a plasma viral load that was between 31 and 85 times that of the nonalcohol control group. CD4 cell loss was significantly more pronounced in the alcohol group beginning as early as one week after HIV infection [4,5]. Progression from HIV to AIDS comes much quicker when alcohol is involved [6]. The conclusion of the Bagasra team nearly two decades ago has been substantiated ever since: "HIV-1 replication may be augmented by alcohol in HIV-1-infected individuals, and alcohol intake may increase an individual's risk for acquiring HIV-1 infection" [7].

12.3 Alcohol abuse by students

Educational officers at prominent U.S. institutions of higher education have become alarmed at the collective impact of individual student decisions about alcohol. They are putting the matter higher on their already busy agendas. Ohio University, the University of Georgia, the University of Iowa, and the University of California, Santa Barbara (ranked by the Princeton Review as numbers 1, 2, 4, and 5 among US *party schools*) are making serious attempts to counter alcohol abuse, and its related problems, among their students. They, and others who deal with or live among persons who abuse alcohol (virtually everyone in the United States), have abundant reason for alarm. Consider these statistics for the year 2009 from the

"College Drinking—Changing the Culture" website [8]. Death and injury: 1,825 college students, ages 18–24 years died, and an additional 599,000 were unintentionally injured in cases where someone was under the influence of alcohol. Drunk driving: 3,360,000 students drove while under the influence of alcohol (little wonder insurance rates are high for young drivers) [9,10,15]. Academic problems: approximately 25% of students admit that drinking had a negative impact on their academic performance; it influenced their lower grades, they missed classes, poor performance on papers and exams, and falling behind in their studies [11–14]. Assault: nearly seven-tenths of a million students were assaulted by drunkard students [9,10,15]. The list goes on, and includes sexual behaviors known to be directly related to the spread of HIV and AIDS: the number of alcohol-related cases in which 18 to 24 year-old students were victims to sexual assault or date rape episodes was an astounding figure of 97,000, but that number was dwarfed by the 400,000 students who had unprotected sex, and those troubled students (over a tenth of a million of them) who confessed to being so intoxicated that they could not say whether or not they consented to having sexual relations [8,9,10,15]. The probability of disease proliferation obviously soars on the drunken wings of such irrational behavior. Alcoholic cocktails can cause a desperate, life-saving need for retroviral cocktails. One leads to the other—the first cocktail by choice, the second a grasping-at-straws necessity. Yet the immediate cause of HIV infection has generally been the questionable, or blatantly immoral behavior by one or both parties in an intimate relationship.

12.4 Compromised immune capacity

Immune deficiency is, of course, at the heart of AIDS-related deaths, and, according to the WHO, the immune system is compromised by alcohol consumption. This was reported by one of the coauthors of the present study in 1989. HIV replicates more readily in bodies weakened by alcohol consumption. Moreover, the weakened immune system enables pathogenic infections, which in turn leads to higher levels of tuberculosis and pneumonia. With heavy drinking, this becomes even more pronounced [1,16]. For persons with immune systems already compromised by AIDS, alcohol-induced attacks on immunity are particularly troubling. Moreover, Baliunas et al. [17] have reported what seems intuitively obvious: initial infection with HIV is more likely if an individual drinks alcohol. Hendershot et al. [35] have observed that even HIV/AIDS patients who have access to antiretroviral treatments are less likely to benefit from those treatments if they drink because they are less likely to adhere to prescribed routines [1]. A BBC News play on word warned of the link between alcohol and immune deficiency illness: "Alcohol 'aids HIV cell infection'" [18]. Citing a study in the *Journal of Acquired Immune Deficiency Syndromes,* the BBC reported that alcohol predisposes cells in the epithelium (the lining inside the mouth) to be receptive to HIV infection. Following a ten-minute exposure to 4% of ethanol, HIV susceptibility was increased between three and six times. The deadly sequence, simply put, goes like this: HIV attacks CD4+ white blood cells. These infected cells then attach to the mouth's lining, to the cells of the endothelial. "HIV hijacks the cell, inserting its own genes into the cell's DNA and uses it to manufacture more virus particles. These go on to infect other cells" [18]. The process begins a chain reaction that can eventually produce full-blown AIDS, and then full-blown death. From both biological and behavioral perspectives, drinking and AIDS do not mix. Alcohol consumption increases the probability of HIV infection, and then weakens the immune system at a time when HIV-positive individuals desperately need healthy immune responses. Researchers are intrigued by the beneficial impact that the hepatitis-related virus GBV-C has on HIV, but they find no such "co-infection" advantage with alcohol. To the contrary, alcohol kills, HIV kills, and, when combined, the two kill more effectively through synergistic slaying.

To stay free from disease, the body must have a healthy immune system, but alcohol can damage routine immune reactions that provide necessary protection [19–22]. White blood cells (notably CD4+ T cells) are critical in the fight against infection, and while they do not fight HIV effectively, they do fight other infections very well. Not only is HIV not intimidated by CD4+ T cells, but it actually thrives on destroying them. This helps to explain why AIDS patients typically die of something not directly related to the HIV virus itself. The very name of the condition, immunodeficiency, is a key term in the acronym AIDS. Researchers have found that, in laboratory animals, alcohol intake causes a reduction in the total number of white blood cells that are available to counter disease-spreading pathogens [23–25]. Closer to home, they have found that the same holds true for humans [26,27]. Drinking alcohol can artificially limit the production of antibodies, and hold back other normal immune reactions in humans [28,29], and animals [23,24]. Macrophages are a particularly helpful type of immune cell, and, among other things, they prevent lung infections. Like a villain attacking a hero, alcohol suppressed macrophage activity [30,31]. Alcohol consumption can even affect the next generation. Women who drink while pregnant can pass on weakened immune responses to their babies [32–34]. Indirectly, being high on alcohol can indirectly lead to being low on life-saving antibodies.

A study by Bagasra et al. concluded that even one drinking experience could depress white-blood-cell immune responses. This investigation, based on research that used white blood cells from volunteers judged to be in good health, found that when isolated, these cells were more prone to infection by HIV retroviruses than were the cells of subjects who had not taken one or more drinks of alcohol [35].

Alcohol is a drug and, as with drug abuse generally, participants are more likely to use poor judgment, take unwise risks, and become infected with HIV. The National Institute on Drug Abuse (NIDA) explains, "drug abuse treatment is HIV prevention" [36]. A mere glance at the titles of two recent scholarly articles by Baum, et al. suggests reason for concern, a reaction that is reinforced by the details in these articles: "Alcohol Use Accelerates HIV Disease Progression" [37]; and "Crack-cocaine use accelerates HIV disease progression in a cohort of HIV-positive drug users" [38]. Whether one argues that alcohol and drugs negatively affect behavior, and behavior in turn increases the chance of becoming HIV-positive, or whether one maintains that alcohol and drugs speed up the progression from HIV infection to critical immunodeficiency AIDS, the conclusion is the same: (1) alcohol, by itself, adversely affects human health, (2) cocaine use, by itself, adversely affects human health and (3) HIV adversely affects human health. Individually, they attack health; collectively, they assault it more frequently, and more quickly. Abstinence from drugs and alcohol is not only logical from an individual standpoint, but also from a public-policy perspective. HIV statistics would improve, as would many other troubling indicators of societal well-being. It is time for the drunken wings on which disease proliferation soars to be clipped.

References

1. World Health Organization (WHO). Global Status Report on Alcohol and Health 2011.http://www.who.int/substance_abuse/publications/global_alcohol_report/en/, accessed February 25, 2011.
2. World Health Organization (WHO). Call for action to reduce the harmful use of alcohol. http://www.who.int/mediacentre/news/releases/2010/alcohol_20100521/en/index.html, accessed December 12, 2011.
3. Nebehay S. Alcohol kills more than AIDS, TB or violence. *Reuters.* 13 February 2011. http://www.asiaone.com/Health/News/Story/A1Story20110213-263238.html, accessed December 12, 2011.
4. Kumar R. et al. Increased viral replication in simian immunodeficiency virus/simian-HIV-infected macaques with self-administering model of chronic alcohol consumption. *Journal of Acquired Immune Deficiency Syndromes* 2005;39(4):386–390.

5. Poonia B, Nelson S, Bagby GJ, Zhang P, Quniton L, Veazey RS. Chronic alcohol consumption results in higher simian immunodeficiency virus replication in mucosally inoculated rhesus macaques. *AIDS Research and Human Retroviruses* 2006;22(6):589–594.

6. Bagby GJ, Zhang P, Purcell JE, Didier PJ, Nelson S. Chronic binge ethanol consumption accelerates progression of simian immunodeficiency virus disease. *Alcoholism: Clinical and Experimental Research* 2006;30(10):1781–1790.

7. Bagasra O, Kajdacsy-Balla A, Lischner HW, Pomerantz RJ. Alcohol intake increases human immunodeficiency virus type 1 replication in human peripheral blood mononuclear cells. *Journal of Infectious* 1993;167(4):789–797.

8. A Snapshot of Annual High-Risk College Drinking Consequences. College Drinking—Changing the Culture. http://www.collegedrinkingprevention. gov/statssummaries/snapshot.aspx, accessed December 12, 2011.

9. Hingson RW, Heeren T, Zakocs RC, Kopstein A, Wechsler H. Magnitude of alcohol-related mortality and morbidity among U.S. college students ages 18–24. *Journal of Studies on Alcohol* 2002;63(2):136–144.

10. Hingson, R. et al. Magnitude of Alcohol-Related Mortality and Morbidity Among U.S. College Students Ages 18–24: Changes from 1998 to 2001. *Annual Review of Public Health*, 2005;26:259–279.

11. Engs RC, Diebold BA, Hansen DJ. The drinking patterns and problems of a national sample of college students, 1994. *Journal of Alcohol and Drug Education* 1996;41(3):13–33.

12. Presley CA, Meilman PW, Cashin JR. *Alcohol and Drugs on American College Campuses: Use, Consequences, and Perceptions of the Campus Environment.* Vol. IV: 1992–1994. Carbondale, IL: Core Institute, Southern Illinois University, 1996a.

13. Presley CA, Meilman PW, Cashin JR, Lyerla R. *Alcohol and Drugs on American College Campuses: Use, Consequences, and Perceptions of the Campus Environment,* Vol. III: 1991–1993. Carbondale, IL: Core Institute, Southern Illinois University, 1996b.

14. Wechsler H, Lee JE, Kuo M, Seibring M, Nelson TF, Lee HP. Trends in college binge drinking during a period of increased prevention efforts: Findings from four Harvard School of Public Health study surveys, 1993–2001. *Journal of American College Health* 2002;50(5):203–217.

15. Hingson RW, Howland J. Comprehensive community interventions to promote health: Implications for college-age drinking problems. *Journal of Studies on Alcohol Supplement* 2002;14:226–240.

16. Lönnroth K et al. Alcohol use as a risk factor for tuberculosis—a systematic review. *BMC Public Health*, 2008;8:289.

17. Baliunas D et al. Alcohol consumption and risk of incident human immunodeficiency virus infection: A meta-analysis. *International Journal of Public Health*, 2009;55:159–166 (Epub December 1, 2009).
18. *BBC News*. Alcohol aids HIV cell infection. http://news.bbc.co.uk/2/hi/health/4123193.stm, accessed December 13, 2011.
19. National Institute on Alcohol Abuse and Alcoholism (NIAAA). Alcohol Alert, No. 15 PH 311 January 1992. http://pubs.niaaa.nih.gov/publications/aa15.htm, accessed November 12, 2011.
20. MacGregor RR. Alcohol and drugs as co-factors for AIDS. *Advances in Alcohol and Substance Abuse* 1988;7(2):47–71.
21. Plant MA. Alcohol, sex and AIDS. *Alcohol and Alcoholism* 1990;25(2/3):293–301.
22. Pillai R, Watson RR. Response to: Alcohol, Sex and AIDS. *Alcohol and Alcoholism* 1990;25(6):711–713.
23. Tennenbaum JI, Rupert RD, St. Pierre RLG, Greenberger NJ. The effect of chronic alcohol administration on the immune responsiveness of rats. *Journal of Allergy* 1969;44:272–281.
24. Jerrells TR, Marietta CA, Eckardt MJ, Majchrowicz E,Weight FF. Effects of ethanol administration on parameters of immunocompetency in rats. *Journal of Leukocyte Biology* 1986;39(5):499–501.
25. Saad AJ, Jerrells TR. Flow cytometric and immunohistochemical evaluation of ethanol-induced changes in splenic and thymic lymphoid cell populations. *Alcoholism: Clinical and Experimental Research* 1991;15(5):796–803.
26. Liu YK. Effects of alcohol on granulocytes and lymphocytes. *Seminars in Hematology* 1980;17:130–136.
27. McFarland W, Libre EP. Abnormal leukocyte response in alcoholism. *Annals of Internal Medicine* 1963;59:865–877.
28. Gluckman SJ, Dvorak VC, MacGregor RR. Host defenses during prolonged alcohol consumption in a controlled environment. *Archives of Internal Medicine* 1977;137:1539–1543.
29. Mutchnick MG, Lee HH. Impaired lymphocyte proliferative response to mitogen in alcoholic patients: Absence of a relation to liver disease activity. *Alcoholism: Clinical and Experimental Research* 1988;12(1):155–158.
30. Guarneri JJ, Laurenzi GA. Effect of alcohol on the mobilization of alveolar macrophages. *Journal of Laboratory and Clinical Medicine* 1968;72:40–51.
31. Rimland D. Mechanisms of ethanol-induced defects of alveolar macrophage function. *Alcoholism: Clinical and Experimental Research* 1983;8(1):73–76.

32. Redei E, Clark WR, McGivern RF. Alcohol exposure in utero results in diminished T-cell function and alterations in brain corticotropin-releasing factor and ACTH content. *Alcoholism: Clinical and Experimental Research* 1989;13(3):439–443.

33. Johnson S, Knight R, Marmier D.J, and Steele R.W. Immunodeficiency in fetal alcohol syndrome. *Pediatric Research* 1981;15(6):908–911.

34. Ewald SJ. T lymphocyte populations in fetal alcohol syndrome. *Alcoholism: Clinical and Experimental Research* 1989;13(4):485–489.

35. Bagasra O, Kajdacsy-Balla A, Lischner HW. Effects of alcohol ingestion on in vitro susceptibility of peripheral blood mononuclear cells to infection with HIV and of selected T-cell functions. *Alcoholism: Clinical and Experimental Research* 1989;13(5):636–643.

36. National Institute on Drug Abuse (NIDA). HIV/AIDS, Research Report Series, http://www.nida.nih.gov/ResearchReports/HIV/hiv.html, accessed October 7, 2011.

37. Baum MK, Rafie C, Lai S, Sales S, Page JB, Campa A. Alcohol Use Accelerates HIV Disease Progression. *AIDS Research and Human Retroviruses* 2010;26(5):511–518.

38. Baum MK, Rafie C, Lai S, Sales S, Page B, Campa A. Crack-cocaine use accelerates HIV disease progression in a cohort of HIV-positive drug users. *Journal of Acquired Immune Deficiency Syndromes* 2009;50(1):93–99.

13

TB and HIV
A Deadly Combination

SIDA (AIDS in Portuguese)

13.1 Tuberculosis and global mortality

Infection with *Mycobacterium tuberculosis* (*MTB*) has long caused serious health problems for humans. Tuberculosis is caused by bacteria and can be infectious from person to person, involved the appearance of tubercles or nodules, particularly in the lungs. Although it typically attacks the lungs, it can attack other parts of the body as well, including the brain and the kidneys. One can be infected with *MTB* and yet not feel or be ill. Globally, tuberculosis (TB) is the leading specific cause of death for HIV-positive people. TB often lies dormant in individuals in a nonharmful state. This is known as latent TB infection (LTBI). When it is actively attacking bodily tissues, we speak of TB disease, and this is deadly if not dealt with properly. HIV infection facilitates the deadly transition from LTBI to TB by weakening the immune system, and encouraging TB to develop as an opportunistic infection as it enters an active and potentially lethal state. LTBI is very common and more than two billion people, or about one-third of the world's population, have it. Approximately 9.6 million people became ill with the disease in 2014, about one million of them were children; about one-sixth of those 9.6 million people, 1.5 million died because of TB, 140,000 of them were children. Fortunately, TB is both preventable and curable. It spreads through germs that are placed into the breathing space of others through coughing, spitting, or sneezing. For HIV-positive individuals who are coinfected with TB, virtually all of them will die of TB if they receive no medical help. For one in three HIV-positive individuals who die, the specific cause of death is TB. Smoking greatly increased the probability that one will develop TB. The two standard drugs that are used to fight TB are rifampicin and isoniazid. These have been prescribed for decades, and over the years drug-resistant disease strains have developed due to poor quality products and failure to take the medication as prescribed. It is estimated that in the year 2014, about a half-million people (480,000) developed multidrug-resistant tuberculosis (MDR-TB), which indicates that a person's immune system cannot effectively oppose rifampicin and isoniazid. A second line of TB drugs is used to preserve the lives of MDR-TB patients [1].

13.2 How TB can be cured in HIV-infected individuals

The number of lives saved by proper treatment following diagnosis of tuberculosis during the first decade of the twenty-first century was about 43 million. Standard treatment requires taking four antimicrobial medications for six months. Adherence to this regimen works best when patients report to a trained health care worker. Although HIV-positive individuals are 20–30 times as likely to have TB than persons who are HIV-negative, treatment is available. The World Health Organization (WHO) recommends 12 preventative and treatment activities. Treatment for HIV patients are available both for individuals with LTBI to prevent TB from becoming active. Treatment is even more imperative for those whose TB is already active. The consequences are potentially deadly if proper treatment is not given. Achieving a proper blend of antiretroviral drugs (ARVs) and TB medications varies according to patient CD4 counts, the state of their TB, and other aspects of their health profile. The typical treatment for LTBI patients who are also HIV positive is to follow one of the following three treatment plans: (1) take isoniazid (INH) over a nine-month period, (2) take INH and rifapentine for three months, or (3) take either rifampin or rifabutin for four months. In addition to the standard TB drugs, growing evidence indicates that antiretroviral or HAART treatments that help prevent HIV from progressing to acquired immune deficiency virus (AIDS) can also prevent LTBI from progressing to active TB. Patients with active TB are encouraged to promptly begin TB treatment according to guidelines for TB patients who are not human immunodeficiency virus (HIV) positive. ARV treatment is also important, both for its preventative blocking of HIV proliferation and for keeping the immune system from deteriorating, which makes the LTBI-to-TB transition much more likely [2,3].

13.3 Comparing two global killers

TB and HIV are both global killers and the two are mutually reinforcing in support of negative outcomes for patients, with TB being the leading cause of death in AIDS patients. As noted in the foreword, there are unsettling similarities in many of the issues surrounding the two diseases. For example, of the estimated 9 million people that develop TB symptoms annually, nearly one-third are perilously oblivious to the status of their infection. Both infections are also treated with cocktails of antibiotics/antiretrovirals, with two or three drugs given to patients at the same time to avoid resistance. In the case of TB, an initial phase of isoniazid (INH), rifamycin, pyrazinamide (PZA), and ethambutol (EMB) are prescribed for the first two months, followed by a continuation phase of INH and a rifamycin for an additional four months. In the case of HIV, an initial phase of a cocktail of reverse transcriptase plus integrase inhibitors is the norm. However, in HIV-TB coinfected people, the TB treatment is initiated first and after a certain time (according to the patient's age, CD4 count, viral load, and other factors), ARVs are initiated. In both types of infections, adverse effects to drugs occur, and the development of drug-resistant strains is common. Due to effective medications, both infections have become chronic infections that require long-term treatment and careful monitoring. No vaccine for either of the infections yet exists. In both cases, infected persons have difficulty adhering to their regimen of medications, which encourages drug resistance and disease transmission.

However, unlike HIV infection, TB can be cured, which had let many to demand more funding to wipe it out globally. Still, it has never garnered the political influence or attracted the brain power that HIV/AIDS has attracted. Therefore, last year a relatively meager U.S. $674 million was donated to cure TB, which kills three times more people than AIDS does. On the other hand, past year HIV/AIDS research and development received more than U.S. $3 billion from one source alone, The U.S. National Institutes of Health.

TB and HIV differ vastly when it comes to diagnosis. Unlike the anonymous free testing and home-test kits, TB testing demands cumbersome, lengthy, and semi-invasive diagnostic procedures. It requires a long time to culture the slow growing TB germs, or chest x-rays or skin injection tests, among other testing procedures. Persons concerned about their TB status need a rapid, home-based diagnostic kit similar to those for HIV, or a simple kit such as those used for pregnancy testing, or a tube that can tell from a person's saliva among their TB infection status. There are many PCR-based (polymerase chain reaction) kits in development, but producers still have a long way to go. So far, all of these diagnostic kits require a patient to visit a hospital or clinic, which adds to inconvenience and cost. The new PCR-based techniques now take only hours, rather than weeks or months as in the past. Direct progress on the TB front means at least indirect progress on the HIV front. The reverse is also true. Both require scientific zeal and appropriate funding. Resources are limited, but the price of not finding better solutions, if not outright cures, is financially high and even higher in human terms [4].

References

1. Centers for Disease Control and Prevention. Tuberculosis (TB). http://www.cdc.gov/tb/topic/basics/, accessed September 3, 2016.
2. AIDSinfo. Guidelines for the Use of Antiretroviral Agents in HIV-1-Infected Adults and Adolescents. https://aidsinfo.nih.gov/guidelines/html/1/adult-and-adolescent-arv-guidelines/27/tb-hiv, accessed September 3, 2016.
3. World Health Organization. Tuberculosis. http://www.who.int/mediacentre/factsheets/fs104/en/, accessed September 3, 2016.
4. Cohen J. TB community borrowing a page from HIV/AIDS. *Science*, December 8, 2015. http://www.sciencemag.org/news/2015/12/tb-community-borrowing-page-hivaids, accessed September 3, 2016.

14
HIV in the Age of the Desaparecidas

AIDS (AIDS in Lithuanian)

14.1 Abortion and HIV: What does abortion have to do with HIV infection?

This question needs to be asked more often. The moral aspects of the question have been widely debated, but what does abortion have to do with HIV? Abortion is commonly used as an attempt to erase the effects of selfish and irresponsible private behaviors, the same behaviors that vastly increase the probability of human immunodeficiency virus (HIV) infection. Human choices are at the root of both abortion and HIV infection. Even the innocent victims are, ultimately, affected by the unwise decision of some other human, however indirect that decision may be. As troubling as this is from the perspective of private morality and the rights of the unborn, the problems multiply when abortion is linked with gender selection.

14.2 Gendercide in the contemporary world

Typically, there are 105 male births for every 100 female births (because more male infants die than female infants). When this ratio is intact, populations persist, and the advantages of gender balance strengthen civilizations. Opposition is an inherent characteristic of human life, and this is apparent in the abuse of the marvelous capacity that ultrasound technology provides. Ultrasounds have done much to contribute to the health of both mothers and babies. When misused, as is commonly done in the world's two most populous countries, China and India, consequences flow that threaten the well-being of mothers, babies, and entire civilizations. Generally, ultrasounds can reveal the gender of a baby, but when mothers, or others involving persons, use that knowledge to promote the elimination of females, a deviation from the positive potential of technology occurs, and societies suffer [1–6].

China and India have the dubious distinction of leading the world in what has, with no exaggeration, been dubbed *gendercide* (a phrase coined in 1985 by Mary Ann Warren). In spite of weakly enforced Chinese and Indian prohibitions against using ultrasounds to unveil the gender of the unborn, Asian gendercide has become so prevalent that the losses in human life have become truly staggering. Surely, the intent of ultrasound inventors and developers was not to facilitate high-tech poaching against defenseless female fetuses. Journalist Mara Hvistendahl estimates that there are about 163 million Asian women who are not alive today who would be had it not been for a massive number of abortions. This number is comparable to that of all women now living in the United States. If the entire U.S. female population were annihilated due to conscious human choice it would be deemed a disaster of the greatest magnitude, yet the fetus-by-fetus elimination of females in Asia receives only limited attention [1–6].

Ironically, a perverse manifestation of modern feminism has provided encouragement to the chief culprits, women themselves, in their determined war against female births. Their supposed private and collective victories are certainly pyrrhic, and the world is suffering, as it always does, from the effects of war. Nor is the problem strictly an Asian disaster. Azerbaijan, Armenia, Eastern Europe, and the United States (in particular population groups) have also contributed to this pyrrhic feminist victory. The global gender ratio has now been altered because of private decisions made in Asia, and elsewhere. Reviewers of Ms. Hvistendahl's *Unnatural Selection: Choosing Boys Over Girls, and the Consequences of a World Full of Men* [2] debate her interpretations, but they are justly impressed with her research, and the projected impact of her book. On a global stage with a preponderance of Asian actors, *Unnatural Selection* includes a varied cast that includes geneticists, mail-order brides, acquired immune deficiency syndrome (AIDS) researchers, prostitutes, strident nationalists, fertility doctors, and prospective parents [1–6].

This number of missing females is far larger than the total number of AIDS deaths (perhaps five times as numerous). It is far larger than the number of *desaparecidos* (the disappeared ones of both genders, both *desaparecidos* and *desaparecidas*) that died in Argentina's *dirty war* of the 1970s and 1980s. The current attack on female life promotes an ever larger population of *desaparecidas*, a population notable for its absence rather than its presence. It is an absence that propels the probabilities of great civilizations in a negative direction, and such statistical trends are not simply due to extraordinary outliers, as can happen in calculations of wealth. It is an individual's decisions that have contributed to the vast throng of missing females. World War I is commonly blamed for significantly wiping out a generation of young men in France. Russian young men and those of other nations also suffered terribly. These *desaparecidos* caused incalculable heartache, altered demographic trends, and deprived their nations of children, inventions, writings, manpower, and so many other things. The total number of current *desaparecidas*, commonly carried out in the name of female liberty, may be about ten times the number of total deaths in the Great War [1–6].

14.3 Male–female ratios and HIV

Human decisions cause statistics; statistics do not automatically force human decisions. Still, it seems inescapable that the probabilities of violence, sexual promiscuity, HIV infection, and a veritable etcetera of other troubles will result from the *desaparecidas*. Asia is playing major leadership roles in the world today, but gendercide is not an area to be proud of.

14.4 AIDS: Asia's Intolerant Daughter Siege

Historically, the male–female ratio has been disrupted many times, but it also has a tendency to work its way back into the normal range. Unnatural events such as wars, epidemics, and disasters of various kinds have affected the balance, but the pattern of selective abortions is unique to the modern world. In India, the ratio has risen to an abnormal 112. In China it is even more abnormal: 121, but this is just an average. The rate for many Chinese towns has skyrocketed beyond the 150 level. The abnormal ratio of the two Asian population giants is sufficient to skew the average for the entire planet to its present level of 107. Armenia's level is 120, Georgia's 118, and Azerbaijan 115. According to Ms. Hvistendahl, it is the rich who generally lead the way: "Sex selection typically starts with the urban, well-educated stratum of society" [2]. The journalist explains, "Elites are the first to gain access to a new technology, whether MRI scanners, smart phones—or ultrasound machines" [2]. She describes a filter-down process that institutionalizes abuse of technology downward to the poorer classes. She reports that the antifemale abortion trend is primarily promoted by the potential mothers themselves, and at times even the mothers-in-law. In either case, it is a woman-versus-woman problem [1–6].

Ms. Hvistendahl warns that male–female imbalances create societies that can be miserable to live in: "Historically, societies in which men substantially outnumber women are not nice places to live" [2]. She warns of instability and violence, as occurred during the troubled times of Athens in the fourth century B.C., the Taiping Rebellion in China during the nineteenth century.

In both cases, female infanticide was a troubling reality. The instability and violence in the frontier era of U.S. history also gives evidence of the danger of letting male–female ratios get out of hand. In 1870, California had 166 males for every 100 females. The rate was even worse in Nevada (320:100), and incredibly out of balance in Kansas (768:100). In contemporary Asia, increases in crime have followed a rise in the gender gap. In India, income level is a leading indicator of the probability of criminal behavior, but the gender ratio is currently the best known predictor [1–6].

Several Chinese provinces have sex ratio at birth (SRB) levels above 130, including Hainan, in southern China, and Henan, a province in the central part of China. India has normal SRB rates in the states of Andhra Pradesh and Kerala, but in the capital (New Delhi), the northern breadbasket area of Punjab and Gujarat, the SRBs are up to 125. Birth order is a statistically significant factor. Parents are less tolerant of female births as their number of children increases. Data from South Korea show how antifemale activity becomes more severe as male-childless couples have additional children. In 1990, the ratio in South Korea was a normal 104–100. Yet in that same year, the number rose to 113 for second births, soared to 185 for third births, and skyrocketed to 209 for fourth births. Moreover, immigrants to the United States from China, India, and South Korea tend to enact birth selection activities in the New World that mimic the patterns of their ancestral homelands [1–6].

A look at the statistics for unmarried 28- to 49-year-old Chinese suggests untold troubles for China, now and in the future. Only 6% of these single adults are female. Of the 94% who are males, few (only about 3%) have finished high school. Concerns about criminal activity, interpersonal violence, and psychological abnormalities seem well worth taking seriously. So many men will never have the chance for marriage. So much unnatural pressure will affect marriage decisions. So much will be out of balance because of the *desaparecidas*. Unmarried men are more likely, based on historical patterns, to have intimate relations with other men. This will increase the probability of HIV infection in Asia. Prostitution will be more likely, which likewise will increase the rate of sexually transmitted infections. Crime will likely increase as well, as will the number of injecting drug users, a high risk group for developing, and dying of, acquired immune deficiency syndrome (AIDS) [1–6].

What does abortion have to do with HIV infection? The question needs to be asked more often. Nothing is not the answer. In the beginning, there was one Adam and one Eve. Things were in balance those days, but today they are not in this age of ultrasound. Still, ultrasound is not the culprit;

humans are. Much like the common pattern in which older immigrants persecute newer immigrants, too many of today's older females have persecuted newer and unborn females. More Eves, and more Adams, need to write, persuade, speak, preach, legislate, and lead in a way that does not discriminate against women in ways that really matter, and in a way that does not allow the tyranny of some older females to rob newer females of their visas to places like China and India, and their passports to the planet Earth. What does such immigration policy have to do with HIV infection? A lot. HIV/AIDS is, finally, very much related to interpersonal relationships. Even with gender balance, HIV infection is a global pandemic. Gross gender imbalances will only make things worse. A well-known acronym, when applied to gender selection, might well be given double meaning as Acquired Immune Deficiency Syndrome is made even more troubling by Asia's Intolerant Daughter Siege [1–6].

References

1. The Impact of Sex Selection and Abortion in China, India and South Korea. *Science Daily*, March 14, 2011. http://www.sciencedaily.com/releases/2011/03/110314132244.htm, accessed November 21, 2011.
2. Hvistendahl M. *Unnatural Selection: Choosing Boys Over Girls, and the Consequences of a World Full of Men*. New York: Public Affairs, 2011.
3. Douthat R. 160 Million and Counting. Opinion Pages. *New York Times*, June 26, 2011. http://www.nytimes.com/2011/06/27/opinion/27douthat.html?_r=1&partner=rssnyt&emc=rss, accessed November 21, 2011.
4. Last JV. The War Against Girls. *Wall Street Journal*, June 24, 2011. http://online.wsj.com/article/SB10001424052702303657404576361691165631366.html, accessed November 21, 2011.
5. Hvistendahl M. http://marahvistendahl.com/index.php/book/, accessed November 21, 2011.
6. McDermott J. A world full of men. *Expat*. http://my.telegraph.co.uk/expat/tag/mara-hvistendahl/, accessed November 21, 2011.

15
The Military-AIDS Complex
AIDS and International Security

Sida (AIDS in Haitian)

15.1 Health and the military

Part of the expansive human dimension of the acquired immune deficiency syndrome (AIDS) crisis in India relates to national and international security, including that of the United States. Although good general health is important for the general populace, it is critical for military personnel. Very legitimate concerns over the alarming rates of human immunodeficiency virus (HIV) infection among troops in Africa have raised awareness of the need for healthy troops in India. Whether for civilians or military personnel, behavior that is both moral and safe is paramount. An April 11, 2004 CBS News report cited that the Central Intelligence Agency (CIA) concerns over AIDS involve military, economic, and health considerations. India's independence from al-Qaeda in a region plagued with ties to the terrorism organization, plus its strong economic attachment to the United States are reasons to worry policy makers. The CIA report argued that a worsening of the AIDS epidemic in India would jeopardize its military strength by weakening the health of its army. AIDS would also hurt the South Asian nation's economy, which would no doubt send ripples through the American economy, with which it is closely linked. With one of six humans residing in India, and given the extensive travel in and out of India, an HIV-infected India would likely lead to higher levels of infection elsewhere [1].

Over a year after the al-Qaeda's attack on New York's Twin Towers, the then Secretary of State Colin Powell urged the United Nations to recognize the link between AIDS and global stability: "AIDS is more devastating than any terrorist attack, any conflict or any weapon of mass destruction. AIDS can destroy countries and destabilize entire regions" (cited by CBS, April 11, 2004). An important U.S. ally in its drive against terrorism, India is an important American ally [1].

15.2 HIV and national security

George Tenet, the director of the CIA, argued that HIV was a threat to national security that *can diminish military preparedness and further weaken beleaguered states.* The security risk is linked not only to private moral behavior, which obviously has enormous implications vis-à-vis the spread of AIDS, but to economics. ARV drugs, which slow the progress of the disease, must be taken consistently or drug-resistant varieties develop more quickly. Impoverished nations have contributed less to this problem for the tragic reason that their people often have no access to the drugs. India, more prosperous than the beleaguered nations of Africa, has the dual benefit of greater ability to purchase the medications and cheaper prices, due to lax patent laws. This easier access, however, is a two-edged sword: it prolongs lives, but can lead to global drug resistance problem, particularly when patients do not, or cannot, take the medicine as prescribed. India's plight becomes the world's plight, given the interconnectedness of the planet [1].

15.3 India and American security

Dr. Suniti Solomon, who in 1986, first discovered a person infected with the AIDS virus in India, remembered, in 2004, "I used to see one patient, new patient every week in 1991, 1992" and noted the alarming spread of the disease: "Today, we see 10 to 11 new patients every day." She established one of the earliest hospitals to deal with AIDS cases and, unlike other hospitals, sought to treat rather than avoid AIDS patients: "We don't throw them out." According to her, about one in five of the patients she deals with are truckers [1].

In the case of HIV/AIDS, lack of understanding can be deadly: *some truckers think bathing after being with a prostitute will do the trick, and they then go on their way making deliveries across this huge country— delivering, among other things, the HIV virus to other prostitutes, and often to their own wives.* Nine of ten of Dr. Solomon's female patients were infected by their husbands, to whom they were faithful. Solomon lamented the lack of protection available to women who were part of arranged, male dominated marriages [1].

In a complex and interrelated world, the military activities of the many nations influence each other. They also influence the political issues that governments deal with. Political and military decisions receive, and deserve, a good deal of attention, but so do health-related concerns. As noted earlier, the health status of India's troops influences the United

States. This is simply one brief case study from one large nation. The United States and other nations across the globe have a vested interest in the health of the military forces of other nations. This may seem counter-intuitive or even illogical. An outlook governed by malice and envy might wish for epidemics that would undermine foreign troops. Although lacking in goodwill, an attribute much needed in the diplomacy of our era, such an outlook is unwise on both moral and practical grounds. Military personnel interact with civilians, and both military and civilian individuals interact with others outside their boundaries. Good health for one helps preserve good health for all.

15.4 Sexually transmitted infections and the military

Sexually transmitted infection (STI) disease figures are typically higher for military personnel than for civilians. A statement from the Joint United Nations Programme on HIV/AIDS (UNAIDS) is a cause for concern: "During peacetime, STI (sexually transmitted infection) rates among armed forces are generally two to five times higher than in comparable civilian populations; in times of conflicts, they can be more than 50 times higher" [2]. This is one of the numerous reasons why war should be avoided as an instrument of international policy, except where it is clearly warranted on moral grounds. Most soldiers are fairly young, and are in a high-risk age group. The use of children in the military, as is common in Africa, increases the probability of abuse, and of STI diseases, including AIDS. It was hardly encouraging, or inspiring, when The South African National Defence Force announced that rates among its military personnel were only slightly higher than the rate for civilian personnel. With adult civilian infection rates between 15% and 30%, the willingness to disclose infection rates may be commendable, but the infection rates certainly are not. Among the factors that increase the probability are a culture that promotes machismo and risk-taking, easy access to drugs and alcohol, widespread immorality, and fails to discipline senior officials who exploit younger troops, especially children [2].

According to figures for South Africa in 2002, AIDS-related deaths account for a stunning seven in ten (70%) military deaths (123). "Uganda's defence force lost more soldiers to AIDS than to fighting in two decades of war with the Lord's Resistance Army. In Zambia, AIDS-related illnesses have killed more military personnel since 1983 than died in all its military operations combined, including the bloody independence struggle" [2,3].

In *Lessons learned*, the concluding chapter of *The Enemy Within*, book editor Martin Rupiya explains the need for a complex attack on a complex dilemma:

"There is need for a close civil–military approach to tackling the complex HIV/AIDS challenges. The impact of HIV/AIDS on the armed forces has demonstrated the old adage that no sector survives in a vacuum. Owing to the intimate linkages of prevention, transmission, care and treatment, counseling, nutritional support, home-based care, spouses and other partners, no approach will succeed unless it embraces contributions from, and collaboration between, civil society and the military" [3].

AIDS has proven to have great power to destabilize, especially when humans make unstable choices. Destabilization of individuals, couples, families, and nations. Unfortunately, that is part of the legacy of AIDS, and its spinoff: the military–AIDS complex.

References

1. CBS. AIDS Out Of Control In India: Bill Gates Donating Millions to Help Stop Epidemic. April 11, 2004. http://www.cbsnews.com/stories/2004/04/08/60minutes/main610961.shtml, accessed January 18, 2008.
2. HIV/AIDS and the Military. *Plus News*. http://www.irinnews.org/pdf/pn/Plusnews-Media-Fact-file-Military.pdf, accessed December 13, 2011.
3. Rupiya M (Editor). The Enemy Within: Southern African Militaries' Quarter-Century Battle with HIV and AIDS, Institute for Security Studies, 2006, 190–204. http://www.iss.co.za/uploads/FULLPDFENEMYWITHIN.PDF, accessed December 13, 2011.

16
Counting the Costs
More Deadly Than Military Conflict

Abivahendid (AIDS in Estonian)

16.1 Necrometrics

Necrometrics measure the number of dead. Such numbers apply to the numbers of those who die in wars and from diseases. The American Civil War resulted in 620,000 deaths [1], the Spanish Civil War 365,000 [2], and World War I 21,500,000 and World War II about 50,000,000 [3]. Each war caused suffering of troops and civilians. It was primarily caused by men, and fought by men. Men consciously killed men. Men purposefully imprisoned men. Acquired immune deficiency syndrome (AIDS), likewise, kills. It wages war against humanity, and the number of deaths it causes may eventually eclipse the number of deaths in all the previously mentioned wars. The AIDS war may have seen only its beginning battles. Lustful men, in the name of love, inadvertently infect, and self-centeredly slay their male counterparts. They also ignorantly, and sometimes knowingly, infect their wives or female partners. In the early AIDS battles to date, women have also innocently, or guiltily, infected men and newborn children. Genuine love has sometimes facilitated the sowing of death's seeds in wives. Infected mothers, if known precautions are not taken, pass on more than a genetic heritage, and can simultaneously grant life and introduce death.

16.2 Global conflicts

The American Civil War, Spanish Civil War, and the two World Wars of the twentieth century each lasted for a few years, but the war against HIV/AIDS has lasted longer than any of these military conflicts. To paraphrase Sir Winston Churchill, this ongoing war has not reached the beginning of the end, and we cannot even say with certainty that it has reached the end of the beginning [4]. Yet we see no dramatic change in human behavior, and no end to the deadly male-male relationships that launched the early blitzkriegs (German for lightning warfare). Moreover, sensually out-of-control tyrants selfishly spread the war's suffering to women. In search of lustful lebensraum (German for living space, a term used by Adolf Hitler), these morally-challenged Mussolinis selfishly strut onto the stage of public health.

Who-cares apathy breeds appeasement, even as the AIDS Anschluss (German takeover of Austria, without resistance) marches on. Frankly, Franco-like tirades against unnatural lifestyles [5] are excessively extreme and unreasonably lethal. Anti-HIV rhetoric must convert hearts and minds to change behavior, and sometimes partners. More than rolling-thunder [6] funding is needed to bombard human immunodeficiency virus (HIV). Where the bombs are dropped does matter. Nor is any Hiroshima–Nagasaki [7] left-right knockout punch available. Given the absence of a preventive vaccine and the persistent presence of irrational human behavior, the fight between humans and AIDS will go on for many rounds.

References

1. The Price in Blood! Casualties in the Civil War. Civil War Potpourri. http://www.civilwarhome.com/casualties.htm, accessed December 16, 2011.
2. White M. Spanish Civil War. Secondary Wars and Atrocities of the Twentieth Century. http://necrometrics.com/20c300k.htm#Spanish, accessed December 16, 2011.
3. White M. First World War, Second World War. Source List and Detailed Death Tolls for the Primary Megadeaths of the Twentieth Century. http://necrometrics.com/20c5m.htm#Second, accessed 21 December 2011.
4. The Churchill Society. London. http://www.churchill-society-london.org.uk/EndoBegn.html, accessed December 21, 2011.
5. Francisco Franco. Spartacus Educational. http://www.spartacus.schoolnet.co.uk/2WWfranco.htm, accessed December 21, 2011.
6. The Bombing of Hiroshima and Nagasaki. History.com. http://www.history.com/topics/bombing-of-hiroshima-and-nagasaki, accessed December 21, 2011.
7. Operation Rolling Thunder. History.com. http://www.history.com/topics/operation-rolling-thunder, accessed December 21, 2011.

17
AIDS Orphans

Eizu (AIDS in Japanese)

17.1 Orphans and musical insights

As a serious social issue that is global in its impact and yet eminently local and personal, acquired immune deficiency syndrome (AIDS) has inspired music that urges broader perspective regarding a major problem, and greater compassion toward individuals and families. As AIDS extends both its macro and micro reach, music provides one means of helping victims cope, and onlookers understand. Such an understanding is much needed to encourage more caring interaction toward HIV-positive individuals who can safely interact with others in society, with no threat to others. Infected persons can do so much more if the stigma associated with AIDS can be weakened or removed. Jony Jerusalem, herself HIV-positive, wrote "The Aids song, The Stigma song" in response to her personal observations and feelings: "There is nothing I want more fiercely," she explained, "than to fight against and shatter the AIDS social stigma." She turned to music to publicize her thoughts and feelings:

"There are viruses, illnesses, cancers, disease
That can wrestle the strongest of men to his knees
But of all the diseases that cut short one's life
Only AIDS has a stigma that cuts like a knife"

"No, don't turn away from him, don't try to hide
From the person who suffers this sickness inside
No, don't turn away for the sake of a name
For you know, deep inside, we are all just the same" [1].

17.2 Case study: Uganda

Barbara Luke writes of the visit that she and her husband Larry made to Uganda, and how they met with children who not only were HIV-positive but also had lost their parents. The Lukes are the U.S. managers for an organization known as the Child2Youth Foundation, an organization that supports HIV-afflicted children in various places of the African continent, provides uniforms to wear to school, furnishes educational supplies, and instills hope. With appropriate nutrition, financial sponsorship, and medical assistance, the lives of these children can more nearly parallel those of other children, and their lives can even be extended, perhaps for two decades or more. As the Lukes visited the schools that these *AIDS orphans* attended, they were welcomed with song and dance, but were surprised at the nature of this entertainment. Rather than hearing popular and cheerful songs about traditional childhood themes, they heard tunes that focused on the infection that threatened their own lives and dreams. Knowing the importance of children properly caring for themselves, and of their need to cope, teachers use music to teach these things. One group sings, "What have we done to deserve you, AIDS?" The tragedy of their query is deepened by their supporting musical declarations: "You are killing the rich, the poor, the old, all our families, and most of all, the young—the future generation—leaving too many orphans. This is not your land. Go away, AIDS" [2, p. 75].

The Lukes observed other Ugandan children who intoned a morbid song over a little child they had symbolically wrapped in a blanket. The children began to weep; so did the Lukes. Their parents would weep as well, but they have departed into another sphere. Like actors forced to play roles in a tragedy they find objectionable, these dramatic players must contribute to a deadly plot they wish were nonexistent. Mrs. Luke echoed the feelings of so many as she searched for words: "What do you say to little children who feel abandoned and dare not dream of a future?" She assured them of unseen realities: of God, and of parents who live on elsewhere. She reminded them that there are visible beings that also care: "God sent us to tell you how perfectly beautiful you are, and we love you, too." [2, p. 75].

17.3 Psychological trauma

As would be expected, AIDS orphans commonly experience psychological trauma. A study conducted in rural Uganda found psychological distress levels to be high. AIDS orphans were more likely to become angry, feel depressed, and suffer from anxiety. Some 3% of all children involved in the study said they wished they were no longer living; for AIDS orphans, the figure was four times that high. This study indicates that providing physical resources, which are too often lacking, is not enough. Serious social and psychological issues must also be addressed [3].

Psychological stress increases with separation from parents, but when this is coupled with separation from siblings, problems increase further. This is a regular occurrence in some regions. In Zambia, a 2002 survey found that more than half of that nation's orphans did not reside with all of their brothers and sisters [4]. Orphans are less likely to have adequate food, shelter, clothing, educational opportunities, and health care. They are more likely to be part of a household presided over by a female, and such households are disproportionately large. They have more people but fewer wage earners to meet expenses [5]. AIDS orphans are often pressured to make financial contributions to those they live with, which increases the likelihood that they will turn to the streets for food or work. They are more likely to become beggars [6], which has its own risks and impacts on feelings of self-worth and personal safety.

17.4 Stigmatization

Stigmatization continues to be a very real problem. Other children and even adults are not always understanding, and may act out of fear, ignorance, or superstition. Besides serious effects on orphans in terms of fear, anxiety, and rejection, these children may also lose access to common societal privileges. The death of a parent is sometimes accompanied by loss of property rights or denial of legitimate claims to an inheritance. As Pelonomi Letshwiti, a social worker in Botswana working for Childline, explained, "You find that the parents have been productive and have left assets for the children but immediately after their deaths, the relatives squander everything. Those that are left without anything are just being used for the food rations" [7].

Although illogical, AIDS orphans are often judged to be HIV-positive because one or both parents died of AIDS. Irrationality often breeds additional irrationality, and if children are deemed untreatable because of their presumed human immunodeficiency virus (HIV) infection status, then they may be denied medical assistance that should be rightfully theirs [8]. The parallel rise in the number of AIDS, orphans and the growing number of young people involved in child labor is especially troubling [9].

Even with compassion and willingness to help in place, AIDS commonly imposes enormous burdens on families and communities. For African nations that have already suffered from civil strife, natural disasters, or epidemics, dealing with the AIDS pandemic is particularly acute. Caring for orphans is not a new issue in Sub-Saharan Africa. Now, however, families and community institutions have found themselves simply overwhelmed by the burdens AIDS inflicts, including lost income and higher expenses, such as those for medical care and funeral services [10]. As generous as donors have been to establish orphanages to care for the parentless children, these institutions should be regarded as only a stopgap measure, a temporary remedy. Life with other family members, or in foster-care settings, is generally preferable. Families are, after all, the basic institution of civilization, and efforts to approximate the family, where it does not exist fully, are generally better than placement in orphanages [8].

17.5 Estimates of AIDS orphans

Uganda has an estimated 1,200,000 AIDS orphans. This is a startling figure, but Uganda has fewer parentless children than do Tanzania (1,300,000), South Africa (1,900,000), and Nigeria (2,500,000). Most of Uganda's orphans are not parentless because of AIDS, but an estimated 44% are orphans because of AIDS. As terrible as this percentage is, it is lower than that of Kenya, whose 1,200,000 orphans comprise 46% of all orphans. The percentages rise even higher in Zambia (53%), South Africa (56%), Malawi (65%), Lesotho (65%), Swaziland (69%), Zimbabwe (71%), and Botswana (72%). Staggering (Uganda, Kenya, Zambia, and South Africa); more staggering (Malawi and Lesotho); most staggering (Swaziland, Zimbabwe, and Botswana) [8]. The staggering macro trends imposed by AIDS are faced daily at the micro level of the individual AIDS orphan. The statement of a 13-year-old reveals the need for both material resources and human warmth: "My sister is six years old. There are no grown-ups living with us. I need a bathroom tap and clothes and shoes. And water also, inside the house. But especially, somebody to tuck me and my sister in at night-time" [8].

References

1. Jerusalem J. The AIDS song. The Stigma song. http://www.hiv.co.il/aidssong/aidssong1.htm, accessed December 7, 2011.
2. Barbara L. Letters from Uganda: Children with AIDS feel despair. *Mormon Times*, March 5, 2011, p. 5.
3. Atwine B, Cantor-Graae E, Banjunirwe F. Psychological distress among AIDS orphans in rural Uganda. *Social Science and Medicine* 2005;61:555–564.
4. USAID/SCOPE-OVC/FHI. Results of the orphans and vulnerable children head of household baseline survey in four districts in Zambia. http://www.fhi.org/en/HIVAIDS/pub/Archive/OVCDocuments/ovczambia.htm, (accessed December 9, 2011), 2002.

5. Monasch R, Boerma JT. Orphanhood and childcare patterns in Sub-Saharan Africa: An analysis of national surveys from 40 countries. *AIDS* 2004;18(2):S55–S65.

6. Salaam T. Congressional Research Service . AIDS orphans and vulnerable children (OVC): Problems, responses and issues for congress. http://www.law.umaryland.edu/marshall/crsreports/crsdocuments/RL32252112304.pdf, (accessed December 9, 2011), 2005.

7. IRIN News. Botswana: AIDS orphans exploited. http://www.plusnews.org/, (accessed December 9, 2011), 2002.

8. AVERT. AIDS orphans. http://www.avert.org/aids-orphans.htm, accessed December 9, 2011.

9. UNAIDS. Report on the global AIDS epidemic. http://www.unaids.org/en/Dataanalysis/Epidemiology/, accessed December 9, 2011.

10. United Nations Children's Fund. Statement by UNICEF representative Bjorn Ljungqvist, HIV/AIDS orphans survey findings conference, 2003.

SECTION IV
Cultural Differences and AIDS Awareness

18
AIDS and the Hispanic Community

18.1 Disproportionate problems

Hispanics/Latinos in the United States face serious problems due to the HIV/AIDS pandemic. Moreover, the experience of this group, whose members can come from any race, has regional and global implications as well. Although in 2006 they comprised only 15% of the United States population, Hispanics accounted for 17% of new HIV infections (Centers for Disease Control [CDC], 2006). By 2009, they accounted for 16% of the population, but their HIV-infection rate had become even more disproportionate, having risen to 20% (9,400) of all new cases in the United States [1–2]. By 2009 they experienced triple the new infection rate experienced by whites (26.4 to 9.1 per 100,000), and the problem had become so severe that among those in the 35–44 age bracket it was the fourth leading cause of death. Obviously, this is a matter of major public policy concern. Current U.S. immigration laws and regulations make living in families exceptionally challenging for Latino men who have come from Mexico, and other Latin American countries. Stringent antiillegal-immigration laws, notably those recently enacted in Arizona and Alabama, have a strong anti-Mexican and anti-Hispanic tone to them. Such laws reduce the opportunities for quality education among Hispanics, make self-sufficiency challenging, and threaten families. This may account, in part, for the atypically high percentage of men living with HIV/AIDS who contracted the disease through sexual contact with other men. Although heterosexual contact accounts for most new infections in the United States, and in the world generally, such is not the case with U.S. Hispanic men. For this group, the three leading means of transmission are men who have sex with men (MSM), injecting drug use (IDU), and heterosexual relations with high-risk partners. For Hispanic women in the United States, the two leading modes of transmission are high-risk heterosexual contact, and IDU. In 2009, 81% of the 6,000 new HIV infections among Latino males were from these MSM [1–2].

18.2 Imbalanced male–female demographic patterns

The imbalanced male–female demographic patterns in many areas only partially explain the exceptionally high rates of new HIV infections among Latino men. Another cause, typically overlooked, is the lack of circumcision among this group. There is now overwhelming evidence that male circumcision dramatically reduces the possibility of contracting HIV. Unfortunately, the already low circumcision rate for Hispanic males seems to be sliding to even lower levels. This unhealthful behavior threatens not only the lives of the men involved, but also those of their spouses, children, and others [1].

A major category that the CDC tracks that relates to acquired immune deficiency syndrome (AIDS) cases among Hispanics is the country of birth for those diagnosed with AIDS in the United States. In 2006, the CDC reported that most cases of Hispanic AIDS transmission (2,608) did not come from a predominantly Hispanic country at all, but rather from the United States itself. Nearly half of these 2,608 cases came from persons born in Puerto Rico (1,346), which was very close to the number born in Mexico (1,334). There were 814 diagnosed cases for natives of Central and South America, and 145 for those born in Cuba. In each of these groups, except Puerto Rico, most cases were attributed to *male-to-male sexual contact*, followed by *high-risk heterosexual contact*, and then *injection drug use*. Most of the Puerto Rico-born cases were the result of *Injection drug use*, followed by *high-risk heterosexual contact*, and then *male-to-male sexual contact* [3].

The types of behaviors that lead to human immunodeficiency virus (HIV) infection are typically self-defeating and irrational. Alcohol abuse, for example, markedly increases the likelihood of unwise decisions regarding intimacy, which in turn take a terrible toll in terms of HIV/AIDS. Lack of condom use, unsafe activity with multiple partners, refusal to abide by high standards of moral behavior, and failure to get tested are likewise irrational *human choices*. More than self-defeating, they are often other-defeating as well. It is important that HIV seropositivity does not become a contemporary rite of passage for young Hispanics who engage in high-risk behaviors [1].

AIDS transmission is, of course, linked to other social and behavioral factors. According to the CDC, "Surveillance data show higher rates of reported STDs among some minority racial or ethnic groups when compared with rates among whites." In the United States, race and ethnicity clearly correlate with what the CDC calls "more fundamental determinants of health status." Poverty certainly exerts its impact, as do the related areas of access to health care, quality of health care, social atmosphere, drug availability, and behavioral modeling. Admitting the severity of STD rates is important, and acknowledging their link to issues of ethnicity and race is likewise important for those who would come to grips with solutions for prevention (*STDs in Minorities*). Regardless of factors that increase the likelihood of infection, however, it is crucial to remember that these factors to not make decisions; humans do. Poor Hispanics can choose to act rationally regarding private moral behavior. Hispanics whose parents have provided poor models of intimate behavior and alcohol use can choose to depart from these self-destructive models [1].

18.3 Young hispanics

Young Hispanic males are particularly likely to make dangerous choices. Of the new HIV cases in 2006 among Hispanic males, 41% were among those who were 13–29 years of age, 16% higher than the corresponding figures for White, non-Hispanic males (25%) in that category, but 1% lower than the 42% registered by Black, non-Hispanic males. Hispanic young women in that age range comprised 36% of the total, compared to 32% for White, non-Hispanic, and Black, non-Hispanic females. In the 30–39 year age range, the percentage of new HIV infections for both Hispanic males, and White non-Hispanic males stood at 34%, a figure slightly higher than the 31% for Hispanic females, and 32% for White non-Hispanic females. For Black, non-Hispanics, the figures were 26% and 30% for males and females respectively [1].

On a percentage basis, the aggregated figures for 13–39 year-olds were most bleak for Hispanic males (75% of the total Hispanic male cases), followed by Black non-Hispanic males (68%), Hispanic females (67%), and Black non-Hispanic females (62%). On the basis of total numbers (as opposed to percentages), most 13- to 39-year-old seropositive male infections came from Black non-Hispanics (10,930), White non-Hispanics (9,650), and then Hispanics (5,530). The order was the same for 13- to 39-year-old females: (1) Black non-Hispanic females (5,480), (2) White non-Hispanic females (2,110), and (3) Hispanic females (1,540) [1].

For new HIV infections among those in the 40 and older bracket, most cases combined came from Blacks (8,510), followed by Whites (7,820), and then Hispanics (2,650). However, it was White males (6,620) who had the most cases of any of the gender-specific groups. They were followed by Black males (5,190) and Black females (3,320). White females accounted for

1,200 new cases, Hispanic males 1,890, and Hispanic females 760. In this older bracket, Black females (3,320) had more new infections than Hispanic males and females combined (2,650). ("Subpopulation Estimates from the HIV Incidence Surveillance System–United States, 2006") [1,3–18].

References

1. Pace DG. HIV/AIDS and the Hispanic Community: Current Overview and Future Prospects. The Year of Change. National Association of Hispanic and Latino Studies. The authors express appreciation to the National Association of Hispanic and Latino Studies for allowing us to include material which Dr. Pace first presented at the 2009 conference, and then published in the proceedings of that conference, 2009.
2. Centers for Disease Control and Prevention. HIV among Latinos. http://www.cdc.gov/hiv/latinos/index.htm, accessed December 12, 2011.
3. Centers for Disease Control and Prevention. Transmission categories and country of birth of Hispanics/Latinos with AIDS diagnosed in the United States during 2006, Centers for Disease Control. http://www.cdc.gov/hiv/hispanics/, accessed January 20, 2009.
4. Addanki KC, Pace DG, Bagasra O. A Practice for All Seasons: Male Circumcision and the Prevention of HIV Transmission. *Journal of Infection in Developing Countries* 2008;2:328–34.
5. Alanis MC, Lucidi RS. Neonatal Circumcision: A Review of the World's Oldest and Most Controversial Operation. *Obstetrical & Gynecological Survey* 2004;59(5):379–395.
6. Auvert B, Taljaard D, Lagarde E, Sobngwi-Tambekou J, Sitta R, Puren A, et al. Randomized, controlled intervention trial of male circumcision for reduction of HIV infection risk: The ANRS 1265 trial. PLoS Medicine 2005;2(11):e298. Epub 2005 Oct 25. Erratum in: PLoS Med. 2006;3(5):e298.
7. Bailey RC, Moses S, Parker CB, Agot K, Maclean I, Krieger JN, et al. Male circumcision for HIV prevention in young men in Kisumu, Kenya: a randomized controlled trial. *Lancet* 2007;369:643–656.
8. Bongaarts J, Reining P, Way P, Conant F. The relationship between male circumcision and HIV infection in African populations. *AIDS* 1989;3:373–377.
9. Ruiz, Sonia, Jennifer Kates, Claire Oseran Pontius, Henry J. Kaiser Family Foundation, *Key Facts: Latinos and HIV/AIDS*. http://www.kff.org/hivaids/6088-index.cfm, accessed January 9, 2009.
10. Centers for Disease Control (CDC). HIV/AIDS, Hispanics/Latinos. http://www.cdc.gov/hiv/hispanics/index.htm. accessed January 20, 2009.

11. Centers for Disease Control (CDC). STDs in Minorities. http://www. cdc.gov/std/stats98/98pdf/Section9.pdf, accessed January 20, 2009.

12. Donoval BA, Landay AL, Moses S, Agot K, Ndinya-Achola JO, Nyagaya EA, et al. HIV-1 target cells in foreskins of African men with varying histories of sexually transmitted infections. *American Journal of Clinical Pathology* 2006;125:386–391.

13. Drain P, Halperin D, Hughes J, Klausner J, Bailey R. Male circumcision, religion, and infectious diseases: an ecologic analysis of 118 developing countries. *BMC Infectious Diseases* 2006;6:172.

14. Gollaher DL. *Circumcision: A History of the World's Most Controversial Surgery.* pp. 53–72, New York: Basic Books, 2000.

15. Kebaabetswe P, Lockman S, Mogwe S, Mandevu R, Thior I, Essex M. et al. Male circumcision: an acceptable strategy for HIV prevention in Botswana. *Sexually Transmitted Infections* 79 2003;79:214–219.

16. Morris, Brian J. Male Circumcision Guide for Doctors, Parents, Adults & Teens. http://www.circinfo.net/rates_of_circumcision.html, accessed February 5, 2009.

17. Williams BG, Lloyd-Smith JO, Gouws E, Hankins C, Getz WM, Hargrove J, et al. The potential impact of male circumcision on HIV in sub-Saharan Africa. *PLoS Medicine* 2006;3: e262.

18. Subpopulation Estimates from the HIV Incidence Surveillance System— United States, 2006 Morbidity and Mortality Weekly Report (MMWR) September 12, 2008; 57(36):985–989. http://www.cdc.gov/mmwr/ preview/mmwrhtml/mm5736a1.htm, accessed January 20, 2009.

19
AIDS and African Americans

SIDA (AIDS in Spanish)

Troubling HIV statistics: HIV infection is a serious crisis in African American communities in the United States that impacts the lives of men, women and children, particularly newborns. According to the Centers for Disease Control (CDC), African Americans face a larger burden of human immunodeficiency virus (HIV) and acquired immune deficiency syndrome (AIDS) than any other group in the nation. The CDC reports that among African Americans, gay and bisexual men comprise more than 50% of all new HIV diagnoses. The statistics are not as bad as in some places in Africa, but they are, nonetheless, bleak: approximately one in 20 men and one in 48 women will be diagnosed as HIV-positive during their lifetime. Although this grim picture has become a little less grim in recent years as the number of HIV-positive African American women has fallen, yet they still are infected at higher rates than female Caucasians, Hispanics, or any other racial or ethnic group. Their level of HIV-1 infection is 20 times higher than that of Caucasian women. Although they make up only 12% of the overall population of the United States, about 44% of the new diagnoses in 2014 were for African Americans. Roughly three-fourths (73%) of these new diagnoses were for males and one-fourth (26%) for females. African American men who have chosen a bisexual or gay lifestyle accounted for most of the new male cases (57%). Of these new male cases, more than one-third (39%) came from youth in their teenage years and early twenties (ages 13–24). During the ten-year period beginning in 2005, there was a 22% increase in HIV diagnosis among all African American men who chose a gay or bisexual lifestyle. During that same 2005–2014 period, there was an even more startling increase of 87% for the gay or bisexual youth (ages 13–24). The correlation between this lifestyle and HIV infection was striking among African American young men living in 21 large U.S. cities. More than one-fourth (28%) of all gay or bisexual young men in 2008 were HIV positive, far higher than the 16% rate for Caucasians who chose that lifestyle. As troubling as this 28% figure is, it is of further concern that three in five (60%) did not even know they were infected, a predictable recipe for HIV proliferation [1].

19.1 Knowledge of serostatus, testing, and the Black Church

The CDC reported that in 2010, nearly 85,000 HIV-positive Black Americans did not know whether or not they were infected. Various factors contribute to this dangerous lack of knowledge, including the initial lack of symptoms that typically follows infection, failure to be tested for HIV, and lack of awareness of risks. The already high percentage of HIV-seropositive African Americans is compounded by the tendency of African Americans to have intimate relations within their racial group. Lack of knowledge of one's HIV status does not, of course, prevent transmission of the infection to others; rather, it makes it more probable. The safest way to avoid this dilemma of not knowing one's status is to choose behaviors that make infection impossible. Given present realities, however, the U.S. Preventive Services Task Force (USPSTF) recommends universal testing for individuals during their 50 years span of life from ages 15 to 65. Universal testing helps remove the stigma associated with voluntary testing, helps those who suspect they may be HIV positive learn their true serostatus, and informs unsuspecting partners who have acquired the infection from someone who has not been tested or have not disclosed their diagnosed condition. Such universal testing would save lives and, given the high costs associated with antiretroviral therapy, would save large amounts of money. An estimated 30% reduction in infection rates would occur if routine HIV testing were implemented. Pinchon and Powell have recommended that the Historically Black Churches assume an even more prominent role in this testing [2]. The recommendation makes good sense given the fact that religion is very important to Black Americans; 85% according to research carried out by the Pew Forum on Religion and Public Life. The church is a place where announcements can be made in person, where attendance at testing events can be encouraged, where events can be organized and coordinated, where volunteer labor is readily available, and where moral values are taught that, if followed, would prevent HIV at the outset. With more than 50% of Black Americans in attendance at least weekly at a church, further focus on the preventative role of the Black Churches seems warranted [2].

19.2 Sexually transmitted diseases and HIV

African American congregations and the communities from which they draw their membership continue to have sexually transmitted disease (STD) rates higher than those of other American racial groups. These STDs (or sexually transmitted infections, STIs), including syphilis, chlamydia, and gonorrhea, are serious in their own right, but they also significantly increase the likelihood of HIV transmission. The same behaviors contribute to all these diseases but, in addition, STDs effect changes in the body that make HIV infection more probable. Every type of STD causes swelling to occur and this inflammation increases HIV risk. Inflammation invites an immune response, and the resulting increase in activated CD4 cells makes it easier for HIV to proliferate, because it preys on these cells. HIV spreads more easily through activated CD4 cells than through those that are not activated. STDs can also cause ulcers, which open additional pathways for HIV entry. Infected persons with STDs, as with HIV-positive individuals, commonly go through a period in which they are asymptomatic, which encourages their belief that they have no infection and this belief is then easily shared by their partners [3,4].

19.3 Late diagnosis

Unawareness of HIV infection takes its toll on both HIV prognosis and the survival rates of the infected. As we mentioned previously, HIV is nearly 100% fatal if untreated. A condition that is untreated, or diagnosed late, typically leads to a number of serious consequences, including an elevated transmission rate to intimate partners, a breakdown of immune processes, and the addition of other diseases that thrive in the weakening of the immune system. In African American communities, tragically delayed diagnosis of HIV infection is common, which results in missed opportunities to receive early medical care and prevent HIV transmission to others [5].

19.4 Poverty, incarceration, and HIV

African American communities experience an elevated poverty rate compared to other racial groups in the United States. This leads to higher levels of other problems, including imprisonment. Lack of resources makes it more challenging to pay for quality health care, to deal with health-related transportation issues, to live in adequate and health-promoting housing, and to pay for HIV medication. These factors explain why viral suppression is a serious problem in the African American community, which increases the likelihood of transmission to others. Those living in impoverished urban areas were as likely to be HIV positive than those living outside urban areas [6].

Incarceration has become a huge business in the United States as the number of inmates has soared in recent decades. In 1980, about a half-million people were incarcerated in American prisons. In 2008, about three decades later, the number had soared to 2.3 million. Although the United States is home to only 1 in 20 of the world's population, 25% of all prisoners are in the United States. About one million African Americans are incarcerated, a rate about six times higher than that for white Americans. When Hispanics are added to the African American totals, the two groups make up 58% of the total prison population. They account for only about 25% of the total population. In 2001, for every 6 black men in America, one had already been behind bars. The National Association for the Advancement of Colored People (NAACP) estimates that this percentage could become far worse, 1 in 3, if present trends are not altered. The rate for African American women is much lower, but still stands at a troubling 1 in 100. Poverty, drugs, and the punitive nature of public policies all contribute to the statistics, and yet with two in three returning to prison, there are serious concerns about whether the annual expenditures of roughly U.S. $70 billion dollars are appropriate. Taxes pay for prisons, but prisons also keep potential workers from working and paying taxes. Prisons breed disease, including HIV/AIDS, and groups that are most prevalent in the prison population also have highest HIV prevalence rates. Attempts to wisely reduce the number of prisoners would no doubt also result in lower levels of HIV and other STIs [7,8].

High incarceration rates are related to high HIV/AIDS rates. Among the major factors in the high rate of HIV infection among African American men, and that fuel multiple sexual partnerships in the African American community, are the large number of sexual relations outside of marriage, low male-to-female sex ratio, high jobless rate, economic oppression, and racial discrimination. All of these are made worse by the astronomically high incarceration rate of African American men. An alarmingly shortage of males due to incarceration nearly doubles the probability that men will have multiple intimate partners. Moreover, the violence by black males against other black males, plus the killing of black males by police officers has further decreased the black male population, and further exacerbated the male–female imbalance. *Black Lives Matter* has brought greater popularity to matters related to racial injustice, including the shooting of unarmed black men. However, deaths in which law enforcement officials were involved, as questionable as some have been, do not come close to the number of deaths of black men by other black men. In 2015, U.S. law enforcement personnel were involved in 1,134 deaths. Of these, African American males between 14 and 34 years of age comprised 2% of the total U.S. population but 15% of those who died after encounters with police (approximately 170). As troubling as some police incidents have been, black-against-black violence has been far more worrisome in its consistency and magnitude. The oft-cited need for black role models is worth mentioning here, role models from all ethnic and occupational groups, both in terms of living peaceful lives, and in promoting safe and moral intimate behavior. External institutions certainly do not replace the home as the primary shaper of values, but they can certainly do their part [8–11].

References

1. Centers for Disease Control and Prevention. HIV among African Americans. http://www.cdcgov/hiv/group/racialethnic/africanamericans/, accessed August 22, 2016.
2. Pichon LC, Powell TW. Review of HIV Testing Efforts in Historically Black Churches. *International Journal of Environmental Research and Public Health* 2015;12(6):6016–6026.
3. Canada's source for HIV and hepatitis C information (CATIE). STIs: What role do they play in HIV transmission? http://www.catie.ca/en/pif/spring-2012/stis-what-role-do-they-play-hiv-transmission, accessed September 3, 2016.
4. Centers for Disease Control and Prevention. HIV among African Americans. Internet. http://www.cdc.gov/hiv/group/racialethnic/africanamericans/, accessed August 22, 2016.
5. Lightfoot AF, Taggart T, Woods-Jaeger BA, Riggins L, Jackson MR, Eng E Where is the faith? Using a CBPR approach to propose adaptations to an evidence-based HIV prevention intervention for adolescents in African American faith settings. *Journal of Religion and Health* 2014;53(4):1223–1235.
6. Denning F. Centers for Disease Control and Prevention. Communities in Crisis: Is There a Generalized HIV Epidemic in Impoverished Urban Areas of the United States? Internet. http://www.cdc.gov/hiv/group/poverty.html, accessed September 3, 2016.
7. NAACP. Criminal Justice Fact Sheet. Institute for Criminal Policy Research. World Prison Brief. http://www.prisonstudies.org/highest-to-lowest/prison_population_rate?field_region_taxonomy_tid=All, accessed August 22, 2016.
8. Clottey EN, Scott AJ, Alfonso ML. Grandparent caregiving among rural African Americans in a community in the American South: challenges to health and wellbeing. *Rural Remote Health* 2015;15(3):3313.
9. Brawner BM. A Multilevel Understanding of HIV/AIDS Disease Burden among African American Women. *Journal of Obsteteric, Gynecologic, & Neonatal Nursing* 2014; 43(5): 633–E50. http://www.ncbi.nlm.nih.gov/pmc/articles/PMC4772147/, accessed August 22, 2016.

10. Swaine J, Oliver Laughland, Jamiles Lartey, and Ciara McCarthy. Young black men killed by US police at highest rate in year (2015) of 1,134 deaths. US policing. https://www.theguardian.com/us-news/2015/dec/31/the-counted-police-killings-2015-young-black-men, accessed September 3, 2016.

11. Carter S. Don't compare police shootings to black-on black crime. Chicago Tribune. 7 July 2016. http://www.chicagotribune.com/news/opinion/commentary/ct-philando-castile-alton-sterling-police-shootings-black-men-20160707-story.html, accessed September 3, 2016.

20
Culture and AIDS Transmission
The Example of India

AIDS (AIDS in Venetian)

20.1 HIV/AIDS statistics

The National AIDS Control Organization (NACO) and other institutions give estimates regarding HIV/AIDS statistics for India, a country in which more than one billion people now live. The results, for 2006 include a major adjustment of an earlier exaggerated estimate of 5.2 million people living with HIV/AIDS. The estimated prevalence is about 0.36%, or between 2.0 and 3.1 million individuals. NACO uses an average figure of roughly 2.5 million. Although that is only about 50% of the earlier estimate of 5.2 million, it is still serious [1]. India's prevalence rate is not nearly as high as some African countries, such as South Africa, which had a 2006 prevalence rate of about 29.1% (down to about 28.0% by 2007) [2]. Nevertheless, due to India's vast population, it is critical to prevent India from ever reaching a threshold from which it would be difficult to prevent a rapid upward spiral of new cases. We have traveled multiple times to India, have made presentations about HIV/AIDS, and have come to some conclusions based on our research, conversations, and observations. These conclusions are very much culturally based. Human immunodeficiency virus (HIV) infection is integrally related to cultural and behavioral patterns. India, with its exceptional cultural diversity, offers an excellent case in point.

20.2 More than a medical problem

HIV/AIDS is not only a biological and medical problem, but also a serious communication challenge made more difficult by the complex cultural patterns of India. Geography, linguistics, religion, and private behavior are all closely linked to culture. India's cultural diversity is both enriching and challenging. Only a culturally-sensitive approach to fighting HIV/AIDS has the potential for cutting through the cultural complexities in a way that is understood by the Indian population. Indian culture often mitigates against coordinated intervention by governmental, nongovernmental organizations (NGOs), and educational institutions, all of which are needed in tackling the interrelated challenges posed by those most responsible for spreading the HIV virus, including truck drivers (noted for their promiscuity), female sex workers (FSW; an obvious high-risk group), men who have sex with men (MSM; easily the highest risk group in India), and intravenous drug users (IDU; infected through unsanitary use of needles). Categories help provide focus, but there is subjective decision making beneath the apparently objective statistical summaries. For example, of the 2006 total of adult acquired immune deficiency syndrome (AIDS) infections in India, 86.2% came from the general population, and the rest were from four high-risk groups: MSM (6.7%), truckers (3.6%), FSW (2.8%), and IDU (0.7%). What if additional high-risk categories could be added, such as those who act under the influence of alcohol were a risk group, those with no strong affiliation with a religious institution, those who live in poverty, those who have intimate relations outside of marriage, or those with very limited formal education? Granted that data collection for these categories would be challenging, but the point we make is that the 86.2% figure for the general population may mask more than it clarifies (1::112).

20.3 AIDS' Achilles' heels

There are several so-called Achilles' heels that may be key culprits in the spread of HIV in India. These include: a truck drivers' culture that encourages high-risk and immoral behavior while on the road, the mistreatment of widows in Hindu culture, the Devadasi system, the absence of circumcision in both Hindu and Sikh cultures, the widespread use of improperly sterilized needles in hospitals, clinics, and physicians' offices, the nature of HIV/AIDS prevention efforts by the Indian government and NGOs, and systemic mistreatment of women in ways that spreads the epidemic. Our travel, contacts, and research have enriched our knowledge about cultural, social, economic, and religious customs in India, and suggested potentially successful preventive strategies that can stymie the spread of HIV.

D. Truck drivers: In India, the most common source of material goods transport and delivery is small trucks. Still, India lacks the infrastructure of the United States and other western nations in which large superhighways accommodate transportation of food and material supplies at high speeds. Small trucks are very efficient in carrying out this function, but often are not allowed to travel in the daytime because of heavy traffic and relatively small roads that are mostly unpaved. Although this no doubt reduces daytime vehicular accidents, it indirectly facilitates greater promiscuity, and the accompanying spread of HIV/AIDS. For religious reasons, cows and bulls are commonly allowed to block the small roads that connect India's villages, and even the congested roads of the cities. Truck drivers generally spend their days in truck stops (dhabas) that generate significant income for local economies. These are not the kind of truck stops we see in the United States, but are small, makeshift sites where drivers stop in the daytime. Truckers find small shops that provide barbers, truck repair personnel, food vendors, and other services. Unfortunately, these stops also have sex workers who exchange their services for money. These are not professional sex workers, but often farm workers who carry out this role to supplement their incomes during the off-season or other times when off-work.

There is now clear evidence that male circumcision (MC) significantly reduces female-to-male transmission of HIV. (We also maintain that MC reduces male-to-female transmission.) In the first randomized controlled trial (RCT) to report on MC, Auvert and colleagues [3] have shown that MC reduces transmission from women to men by 60% (32%–76%; unless otherwise stated, ranges are 95% CI). In sub-Saharan Africa, estimates of HIV prevalence are significantly associated with the estimated prevalence of MC. In countries where fewer than 30% of men are circumcised, the median prevalence of HIV is 17%; where more than 90% of men are circumcised, it is only 2.9%. We maintain that the Indian masses must be convinced, regardless of religious perspective, that it would be greatly beneficial for them to practice circumcision. Our preliminary data show that Africa, with high rates of HIV/AIDS prevalence, still had varying infection rates, with the lowest rates in the areas where a majority of the population practice circumcision (Muslims), whereas the highest number infected was in the uncircumcised groups. India echoes this pattern, with the majority of its infected population coming from the southern region, an area with a low percentage of Muslims. Regions with high percentages of Muslims were observed to have less than 0.1% of their population infected with HIV [3–5].

References

1. National Institute of Medical Statistics, National AIDS Control Organisation, Technical Report: India HIV Estimates-2006. http://www. unaids.org/en/dataanalysis/epidemiology/countryestimationreports/ india_hiv_estimates_report_2006_en.pdf, accessed November 4, 2011.

2. National Department of Health, South Africa. The National HIV and Syphilis Prevalence Survey: South Africa 2007. 2008. UNAIDS. http://www.unaids.org/en/dataanalysis/epidemiology/countryesti-mationreports/20080904_southafrica_anc_2008_en.pdf, accessed November 2011.

3. Auvert B, Taljaard D, Lagarde E, Sobngwi-Tambekou J, Sitta R, Puren A, et al. Randomized, controlled intervention trial of male circumcision for reduction of HIV infection risk: The ANRS 1265 trial. *PLoS Med* 2005;2(11):e298.

4. Jayaprakash Institute of Social Change. Summary Report of the Situation Analysis of Widows in Religious Places of Wet Bengal. New Delhi, India: Ministry of Women and Child Development, Government of India, 2009. http://wcd.nic.in/research/situanwidowswb.pdf, accessed December 13, 2011.

5. Williams BG, Lloyd-Smith JO, Gouws E, Hankins C, Getz WM, et al. The potential impact of male circumcision on HIV in Sub-Saharan Africa. *PLoS Med* 2006;3(7):e262. doi:10.1371/journal.pmed.0030262.

21
Indian Traditions, Women, and HIV

CIDA (AIDS in Venda)

21.1 Disproportionate burdens

Women, as well as children (especially orphans), bear disproportionate burdens in contemporary India. Issues related to the exploitation of women also commonly have application to children. The female portion of the population has the least power to deal with the numerous issues that HIV/AIDS has inflicted upon them. The limitations imposed by economic hardship and prejudicial cultural views about women are not only demeaning to women, but also have an impact on the spread of disease. Although much progress has been made to curtail the oppressive demands of the dowry tradition, a tradition upheld financially by the bride's parents, this unfortunate social convention continues to have an impact on marriage opportunities, and marital tension. The lack of a dowry, or low amount of a dowry, makes marriage less likely, and increases the likelihood for promiscuity, with its attendant health risks. Increased potential for education has resulted in improved conditions for many women, but still vast numbers of women are inadequately educated. Moreover, cultural taboos prevent women from having the voice that is needed in sounding the alarm and teaching others, including other women, about the myths and realities of HIV/AIDS. We have observed that the large majority of those with whom we have dealt in higher education are males, whereas neither higher education nor society, in general, can fight human immunodeficiency virus (HIV) adequately without the serious and pervasive involvement of women. Given the subordinate role in which women and children find themselves in India, we argue that males must shoulder the primary responsibility (and accept most of the blame) for modeling moral and ethical behaviors that are compatible with the best of the various religious and social traditions in India, behaviors that lie at the heart of HIV/AIDS prevention.

21.2 Widows and remarriage

In Hindu culture, once a woman from a lower cast becomes a widow, she is not allowed to remarry and often goes to live in an *ashram*, where she lives an ascetic life along with other widows. This inability to remarry, although not unique to Hinduism or to India, seems to have an adverse impact on public health. The residents of the *ashrams* commonly live in poor and destitute conditions for the rest of their lives, and many become targets of wealthy men who exploit their vulnerability for selfish immoral purposes. According to the 2000 census report, there were 3.4 million Indian widows living in *ashrams*, many near a small or a midsize city. India's Muslim population is the second largest in the world [1]. Islam allows widows to remarry, which alleviates many of the problems that widows face in India.

As discussed in one recent study, *Widowhood in India is a complex institution fraught with contradictions in meaning and practice*. Particularly among the Hindu majority, widowhood isolates women from society, deprives them of normal social and economic privileges, increases potential for loneliness, decreases the likelihood of receiving proper health care, reduces their income, imposes social stigmas, blocks them from future marriage opportunities, and imposes restrictive expectations regarding food and clothing. "Perhaps more than any other social institution in India, widowhood exposes the gap between cultural and social realities, between precept, and practice" [1]. Granted, widowhood poses challenges to the husbandless women, and to society, in the best of circumstances. But conditions in India are particularly perplexing, and widows are particularly vulnerable. The vast numbers of widows in rural areas find themselves incapable of managing crop and land issues without male help. Moreover, the medical debts associated with their now deceased husbands become their responsibilities. Unable to meet the rigors of rural life, they feel pressured to migrate to urban areas, and yet this migration comes with its own risks "of trafficking, physical, and sexual assaults, and so on and many young widows end up in the brothels" [1]. As with avaricious funeral home services in the United States, where the next of kin of the deceased are pressured into paying exorbitant prices for caskets, and then left with thousands of dollars of loans for burials, there are specialized groups in India that lure the newly widowed into entering a new *life* full of promises and hopes. In short, there is, in contemporary India, a troubling correlation between widowhood and acquired immune deficiency syndrome (AIDS). The conditions that challenge millions of Indian widows increase the probability of HIV infection in the world's second most populous country.

21.3 Devadasi and jogini

Devadasi is a religious practice in some parts of southern India, including Andhra Pradesh and Telangana, whereby parents marry a daughter to a deity or a temple. The marriage usually occurs before the girl reaches puberty and requires her to provide sex to upper-caste community members. Whether this lifestyle is viewed as prostitution or religious zeal, it prevents these *jogini* from entering into a real marriage. Mythically, the Devadasis are the incarnation of Urvashi, the celestial nymph. *Joginis* are recognized by their copper bangles, the band they wear around their necks with a leather pendant, and a long necklace with several pendants which have the image of the goddess Yellamma. The practice was legal in much of India until 1988, yet there are still instances of the practice due to continuing superstition, slowness to accept societal reform, and the unwillingness of authorities to intervene. This type of religious prostitution is known as *basivi* in Karnataka, and *matangi* in Maharashtra. It is also known as *venkatasani, nailis, muralis,* and *theradiyan.* This practice gives evidence of the link between religion, superstition, exploitation, and HIV transmission.

21.4 Progressive legislation

We recommend an aggressive program for the eradication of the *jogini* system, which is a remnant of the Devadasi. It is a heinous practice that is often thrust on the poor, *untouchable* women in the remote villages. Large-scale TV advertisements, as well as counseling programs, would help change the attitudes of the *joginis* as well as of the villagers. The enactment of the Andhra Pradesh Devadasi (Prohibition of Dedication) Act of 1988 is commendable:

> "Whereas the practice of dedicating Women as Devadasis to Hindu dieties (deities), Idols, objects of worship, temples and other religious institutions or places of worship exists in certain parts of the State of Andhra Pradesh; and

> "Whereas such practice, however ancient and pure in its origin, leads many of the women so dedicated to degradation and to evils like prostitution; and

> "Whereas it is necessary to put an end to the practice" [2].

The long-overdue legislative enactment in Andhra Pradesh was preceded four decades earlier by a similar measure in Tamil Nadu [3]. Both laws deal with the issue of marriage, and make it illegal to prevent a previous

Devadasi from marriage: "any woman so dedicated (to the service of a Hindu deity, idol, object of worship, temple or other religious institution) shall not thereby be deemed to have become incapable of entering into a valid marriage" [4].

However commendable the wording of the 1947 and 1988 Acts for Tamil Nadu and Andhra Pradesh may be, they also need to be enforced on a continuing basis, widely publicized, and imitated in order to bring greater hope to impoverished poor women, to heap greater shame on those who use social and religious pretexts to prey on others, and to fight AIDS. Although TV holds great promise for the dissemination of accurate and helpful information regarding HIV/AIDS, it also has much negative potential. Indian culture is increasingly influenced by television. Western influences too often encourage the very behaviors that breed disease. We maintain that government has the right and obligation to regulate the media where matters of public health are concerned. We strongly recommend that the religions of the world be used to further the practices that ennoble both males and females, and promote the public good. Acting in accordance with harmful and uncivilized superstitions, regardless of how ancient or enshrined in practice, should not be allowed to exist legally in the civilizations of the modern world, including the great and ancient civilization of India.

References

1. Jayaprakash Institute of Social Change. Summary Report of the Situation Analysis of Widows in Religious Places of West Bengal. New Delhi, India: Ministry of Women and Child Development, Government of India, 2009. http://wcd.nic.in/research/situanwidowswb.pdf, accessed December 13, 2011.
2. The Andhra Pradesh Devadasis (Prohibition of Dedicated) Act, 1988. Laws of India: A Project of PRS Legislative Research. http://www.lawsofindia.org/statelaw/2693/TheAndhraPradeshDevadasis ProhibitionofDedicatedAct1988.html, accessed December 13, 2011.
3. The Tamil Nadu Devadasis (Prevention of Dedication) Act, 1947. Laws of India: A Project of PRS Legislative Research. http://www.lawsofindia. org/pdf/tamil_nadu/1947/1947TN31.pdf, accessed December 13, 2011.
4. Shiva Kumar, ND. Finally, an end to devadasi system. Jan 23, 2009. http://timesofindia.indiatimes.com/city/hubballi/Finally-an-end-to-devadasi-system/articleshow/4023672.cms, accessed December 22, 2016.

22
Confession and Complaint
Bad, Worse, and Worst

ādz (AIDS in Turkish)

22.1 Needle abuse

In her perceptive little volume about confession and literature, Spanish author María Zambrano argues that crises provoke confessions. Confessions are evidence that humans are trying to find themselves, particularly when weighed down with humiliating burdens, or by personal failures. Confession doubles as a means of coping and escape. The despair that inspires confession may spring from guilt, but it is also, like the anguished confession of the Old Testament sufferer Job, *queja, simple queja* (complaint, simple complaint) [1]. The complaint of a confessant is accompanied by hope, a belief that wholeness may yet be possible. Confession is a literary, and living, genre of angst and hopefulness of death of the old life and birth of the new [2]. The confession-complaint of individuals in India about abuse of needles through overuse is a case in point. Over the past decade, researchers have sounded a collective alarm about various needle-use practices in India. In 2000 [3], Singh, Dwivedi, Sood, and Wali wrote about *the reuse of needles by local, medical, and paramedical practitioners for administering antileishmanial drugs*, and warned that "this trend, if not checked immediately, may have drastic consequences in the horizontal transmission of HIV in Bihar." Two years later, Sood, Midha, and Awashthi [4] spoke out against the *sizeable proportion of family physicians in the Punjab state* that practice needle and syringe reuse, and concluded that "information on the virology, clinical presentation, diagnostic tests and management approaches were lacking among a substantial proportion of family physician(s)." In a 2003 study [5] of needlestick injuries, Wig found that "most of the needlestick injuries were neither reported nor investigated." The likelihood of such injuries obviously increases when the number of prescribed injections increases. According to a survey conducted by Murhekar, Rao, Ghosal, and Sehgal in

2005 [6], an injection was part of 18.8% of all prescriptions studied, and the average number of annual injections per capita was three. Consequently, they recommended *remedial measures, such as education of prescribers to reduce the number of injections to a bare minimum.* In a 2005 [7] report on a six-year study of needlestick injuries sustained by health care workers in Mumbai, Rodrigues, Ghag, Bavia, Shenai, and Dastur observed that 380 injuries resulted in 50 cases of testing positive for Hepatitis B, Hepatitis C, or HIV (15 cases). The following year in 2006, Gupta and Boojh [8] published an alarming report on the sloppy and dangerous biomedical waste disposal issues at Balrampur Hospital, which they call *a premier health care establishment in Lucknow.* These authors recommend that not only Balrampur, but also other health care facilities improve their biomedical waste disposal procedures. We join with these researchers in calling for increased attention to needle use, or more specifically, reuse.

The confession-complaint combination is also apparent in an interview with a physician (referred to anonymously as Dr. H.) who was instrumental in persuading sex workers in Hyderabad and Pakistan to transform their lives and educate others about the devastation immorality wreaks on both the human body and the human character. The interview, summarized by Dr. Bagasra and later edited by Dr. Pace, took place on 13 April 2011 at the National AIDS Control Center (NACC) in Karachi, Pakistan after an ad hoc meeting of the NACC earlier that day.

The confession-complaint of this sex worker from Hyderabad, Pakistan is both poetic and pathetic: "I am like a fire hydrant where every dog comes to relieve himself." Dr. H's rapport with these workers allowed her to talk with them, and perhaps even more important, to listen to them, to be their confessor as it were. Her care for these women allowed her to become an understanding teacher to them, one whose goal was to lift them to a higher and safer plane. Her concern to protect these women from self-serving *dogs* seems to have brought greater understanding and safety. Confession and change are common in every nation and every culture; they have a place in the quest to prevent human immunodeficiency virus (HIV) infection. Change is possible, and desirable, even if it comes in small packets. Although communities must take measures to preserve the innocence of the innocent, they must also actively urge change; they must urgently promote the dismantling of risky behaviors and hazardous institutions.

Dr. H's experience: Dr. H., a beautiful and remarkable young Pakistani physician from a conservative Muslim home, became a sex worker's confidant and educator. Her desire to curb the increasing devastation inflicted by HIV/AIDS on this chaotic Islamic nation superseded all her reservations and fears. Intent on stopping the downward slide she determined to do something about her dream. "I came from highly educated Syed home (i.e., a family that claims ancestry back to the prophet Muhammad). My father was minister of education for many years in the Sindh (Indus) Providence (of Pakistan). Two of (my) sisters are also physicians; one resides in Toronto, Canada. After medical school, I wanted to work as a physician, and somehow I decided to work in the red district." In the interview with Dr. H., this young physician recounted the genesis of her desire to make a difference in her society. She confided in her a supportive father, also a determined proponent of reform, who helped her locate a small room, and set up a clinic, in the heart of the red district of Hyderabad. She recalled how she initially decided to work in an area where Hindu leather workers had settled. These middle-class craftsmen, traditionally proud of their profession, were located adjacent to sex workers, called *Khothas* (a term that denotes a small mansion, but in practice is generally used as a word for brothels).

Dr. H. has rather comprehensive insights into the sex business in Hyderabad, the second largest city in the Sindh province of Pakistan, and the fourth largest city in the country. The city is home to about two million people from all walks of life: (1) Sindhi (the native folk), (2) Pathans, (3) Muhajirs (immigrants from former British India), and (4) Punjabis live in the city. The majority of the sex workers in the Sindh province belong to this latter group. The city of Hyderabad was founded in 1768, far earlier than the creation of Pakistan in 1947. Known prior to the 1947 Pakistan–India partition as the *Paris of India*, the city is noted being perhaps the most colorful city of the entire nation, a city rich in industries and craft production. The colorful dresses of the women, regardless of wealth or status, have become a distinguishing characteristic of this urban center.

In this fascinating Pakistani city, Dr. H. made the neighbors aware that she was a doctor, and that she could treat everyone for free, something she could do because of funding received from a government grant designed to help reduce sexually transmitted diseases (STDs), including HIV/AIDS, particularly in sex workers. Initially, she predicted that no one would come. For days, only rarely did anyone enter her clinic, but even these were not the people she was trying to reach. Then slowly, slowly, little children, mostly boys of a tender age, could come and chat with her; these were the children of the red district. Afterward, some older ladies and madams started to come. They inquired about her motives: *Why would you treat us for free?* They questioned whether she was endeavoring to extract HIV-related information from them in order to expose them to the media, particularly the newspapers? They had been bitten by this kind of snake previously, and they were wary of future public exposures. Men had come before to draw blood, to test for disease, and then they announced that HIV-infected workers were among their numbers. There was the shock of the acquired immune deficiency syndrome (AIDS) announcement, and troubling financial repercussions to their business. Word of the infections spread quickly; stigmas, already plentiful in red-district business, became even more negative.

The physician, with a guileless confidence born of pure motives, explained to them that she was not interested in pursuing the paths that had troubled them earlier. She was a doctor who just wanted to help. Early on, she avoided the mention of AIDS, even indirectly. She began treating their children for their fevers; she began assisting mothers with their deliveries. She knew that these unsophisticated common people did not tend to go to doctors but to midwives, and that these women listened to them, heard their confessions as it were. She kept their secrets; she listened to their woman-to-woman complaints. Midwifery runs in families; it is multigenerational with all the benefits to society of a cherished and trusted institution. Unless a midwife recommends that a pregnancy is high risk, none of the sex workers would ever go to hospital or a private OB/GYN office.

In the beginning, Dr. H. would ask a mother for the name of the father of her child. That was a mistake, and she realized it quickly. It was a bad idea to even ask. These children have no fathers in their life. If they have to be hospitalized for some reason, they would use an uncle's name or the name of a male who is closely related to them. Dr. H. slowly gained their trust. She helped them during their times of grief and sorrows, she participated in their rituals, she assisted with difficult deliveries, and she consoled them after miscarriages. She got to know them; she understood their way of life. *These folks are wonderful people*, she commented. "They practice religion, they read (the) Qur'an, observe five-times-a-day prayers, and generally follow (the) Shi'a school of Muslim law. During the past 10 days of Muharram (a holy month for the Shi'a branch of Islam), they close their business and reflect and mourn the death of Imam Hussain (ibn Ali, the Prophet Muhammad's grandson who was assassinated by the Umayyad king). They accept their destiny as sex workers, and do not have remorse." Subsequent actions, when given better options for their lives, revealed that the women did have some deeply-experienced remorse, and were willing to challenge their apparent destiny.

Dr. H. related how the madam initially wanted to know if she wished to close their business, and take their workers away from them, but she was able to convince the madam that she contemplated nothing of the kind. Instead, she began to teach them about STIs, the symptoms of the infections, and how to detect them. Eventually, she started a full-fledged, government-supported HIV/AIDS and STI prevention clinic in the red district. Dr. H. hired women from the same community, a peer-group approach, whom she would train. They would set up education camps, but do so secretly for enhanced impact. Only 10 women were allowed in each group, and the groups were separated into different rooms. Here they were tested for their ability to detect symptoms of the various STIs. Here they were also rewarded, with gifts of their own choice, when they responded with correct diagnoses. They could choose items such as lipstick, talcum powder, or a small make-up kit. The girls were excited! They always wanted to learn new things, longed to expand their horizons. Weary of abuse to their bodies, they also welcomed the opportunity to use their minds.

At the beginning when she was still moving along cautiously, Dr. H. refrained from using the word *condom*. At that time, the use of such a protective device was considered taboo in their business dealings. She knew, and wanted them to know, that such protection could be a matter of life and death for them. These workers had been indoctrinated instead to have unprotected sex with their clients, and to never question their perilous practices. The young doctor made them aware of the enormous protective benefits of condoms, and did so gradually through patient and friendly persuasion. It was the older generation that resisted change; the younger workers came aboard quickly.

Still, the world of the red district was one of tragic choices, selections that were not between good, better, and best, but between bad, worse, and worst. Dr. H. tried to help them make life-saving decisions in their life-threatening occupation. Still, when they found that many of their *dogs* did not want condoms, they resorted to clever methods to get them to use them anyway. Although it is not universal, a high percentage of sex workers and clients use alcohol [9], so the workers learned to give more alcohol to the clients than what they would drink themselves, in order to make them more amenable to condom use. After this, they could use either female condoms or easily slip one-on-one onto the male client. In this world of bad, worse, and worst decisions, some of the workers have desired to become pregnant themselves, and have selected a handsome, fair-skinned, and robust man to father their child. Unlike the overall practice in South Asia and China, they have often preferred female babies over male ones, because they believe females will bring security to their future. They stereotypically consider males to be docile and lame, individuals who irresponsibly depend on the financial support of their females.

Of course, despite Dr. H's personal success, and the satisfaction she feels from knowing she is helping an often helpless and tragic community, she has to pay a personal price in terms of her own reputation. Many times, family or friends have seen her coming in and out of the red district, which prompted them to inquire why a decent person like her, a lady from a conservative Muslim home, would work with such sinners. She has answered them that these unfortunate and abused workers are also God's creations. She has asked them what they might have ended up doing, if they had been raised in similar circumstances, to those whom they now so quickly condemn. They dislike her responses, continue to judge her harshly, and remain unconvinced and unsympathetic to what Dr. H. is trying to do.

The doctor explained that the sex workers in the city can be divided into three major categories: (1) those who work in the brothels, (2) the street hustlers, and (3) the Kotikhana. The brothels are confined to the red district of the city. There the sex workers are commonly Punjabis, and are professionals who form a somewhat caste-like community. The workers in this group are often part of a multigenerational group; their work is a type of unfortunate inheritance. Every female in the group works for the same goal, with older women managing the younger generation. Far from the practices of parents and teachers in most settings, who urge upright behavior, and warn against evil books and media, these women tutor the next generation in the opposite direction, and thus provide them with skills that will unfortunately keep them trapped and miserable, and even lead to their premature deaths. Rather than ignore these individuals, Dr. H. has sought to at least try to improve their plight, and teach them how to protect themselves from the exploitation that characterizes their trade. Protection against HIV in this group also helps to protect the larger society.

Times have changed and techniques have become both more sinister and more sophisticated. Girls often go to designated private places that are preferred by their clients. The street hustlers used to look for business at their own specific corners of the city blocks, but their activities have evolved into a network. These are a totally independent group of sex workers not related to those described above. They are generally educated girls, college or health care industry girls. Yet, in spite of the warnings their education and moral instruction should provide, they promote practices that endanger not only personal and community virtue, but also the health and moral climate of their city and surrounding areas. They can be very sophisticated with their iPhones, and their Internet access. The middlemen they rely on can also be sophisticated in their methods. Although intelligent exploiters of females for personal pecuniary gain, they seek to minimize the level of exploitation of the very women they place in abusive situations. They may even be well-armed, in order to protect their nefarious investment. This group generally avoids any kind of HIV test, or other precautions, that may save their own lives or those of their clients.

In the world of bad, better, and worse we are describing, the third category seems to be the worst. This home-based business is occult, and even more evil than its foul counterparts. A madam operates her exploitative business, for example, by purchasing four to five young girls from Punjab. These become her slaves and her personal property. A madam would pay for these girls according to how she perceives their outward beauty. With demonic skill, she then rents a nice house in a prosperous neighborhood, and sets up the business. The neighborhood is never the same while her business is there, but usually she runs a short-term rental operation, and then moves quickly to other areas if it becomes obvious to the neighbors what she is perpetrating in their neighborhood. Sometimes one of the girls may run away but not for very long. These people are highly connected and will reclaim their property quickly with severe consequences for the girls who seek to escape from their bondage. A headhunter will bring the girl in with satanic zeal. A variation of this home-based approach also mocks the traditional images of a home as a place of love, refuge, and safety. In this variation on a tragic theme, a group of sophisticated, educated, and upper-class girls charge as much as 10,000 rupees per night. Like perverted models, they mask their iniquity with expressive designer clothing, and luxurious cars. These are girls—one would not call them ladies—who sell themselves for money and personal enjoyment. They certainly do not need the money. These delight in sinning with other people of high status including politicians, sports stars, and other wealthy individuals. Drugs, including alcohol, are often part of their world.

In all variations of this devious business, payoffs to law enforcement play an important role. Islam does not promote immorality, and given the predominant place of Islam in Pakistan, such business should, technically, not even exist. This routine *housekeeping expense* threatens homes and health in Pakistan and elsewhere in the South Asian subcontinent. This business, in whatever form, defies the spirit of Islam, Hinduism, Sikhism, and Christianity. It takes a certain type of courage for a Dr. H. to look beyond the corruption, and the moral stench, of the sex business, and seek to protect the lives of individuals who are in virtual bondage. Yet Dr. H. has sought to move those she cares about, and seeks to serve directly, from worst to worse, or even from worse to bad. It is a pathetic world, but Dr. H. sees some reform as better than none, and hopes to do at least something to counter the spread of HIV/AIDS in a largely clandestine environment that encourages immorality, drugs, alcohol, and irresponsibility, the very practices that threaten the virtue and the very lives of private individuals and the body politic.

References

1. Zambrano, María. La confesión: género literario. Madrid: Ediciones Siruela, (1943) 1995.
2. Pace, DG. *Unfettering Confession: Ritualized Performance in Spanish Narrative and Drama.* Lanham, MD, University Press of America, 2009.
3. Singh S, Dwivedi SN, Sood R, Wali JP. Hepatitis B, C and human immunodeficiency virus infections in multiply-injected kala-azar patients in Delhi. *Scandinavian Journal of Infectious Diseases* 2000;32:3–6.
4. Sood A, Midha V, Awasthi G. Hepatitis C—Knowledge and practices among the family physicians. *Tropical Gastroenterology* 2002; 23:198–201.
5. Wig N. HIV: Awareness of management of occupational exposure in health care workers. *Indian Journal of Medical Sciences* 2003; 57:192–198.
6. Murhekar MV, Rao RC, Ghosal SR, Sehgal SC. Assessment of injection-related practices in a tribal community of Andaman and Nicobar Islands, India. *Public Health* 2005;119:655–658.
7. Rodrigues C, Ghag S, Bavi P, Shenai S, Dastur F. (2005) Needlestick injuries in a tertiary care centre in Mumbai, India. *Journal of Hospital Infection* 2005;60:368–373.
8. Gupta S, Boojh R. Report: biomedical waste management practices at Balrampur Hospital, Lucknow, India. *Waste Management & Research* 2006;24:584–591.
9. Samet JH, Pace CA, Cheng DM, Coleman S, Bridden C, Pardesi M, Saggurti N, Raj A. Alcohol use and sex risk behaviors among HIV-infected female sex workers (FSWs) and HIV-infected male clients of FSWs in India. *AIDS and Behavior* 2010;14 Suppl 1:S74–83.

23
The Language of AIDS

AIDES (AIDS in French)

23.1 Linguistic interpretations of AIDS

What affects society affects language, and the global concern over acquired immune deficiency syndrome (AIDS) is no exception. A look at how some African languages describe AIDS gives insights into social impacts of this epidemic. Metaphors say something about human creativity, and patterns of expression, but they also reflect how individuals envision HIV/AIDS. Professor T. Dowling's study of the Cape Town, South Africa area is particularly illuminating in this regard [1]. Owing to its geographic focus, Zulu and Xhosa expressions predominate. In Africa, giving names is routinely done so with meaning in mind. The name *Ntombizodwa*, for instance, means *girls only*, and applies to a child in a family with no male children. A child that comes after a long period of waiting may be named *Lindiwe*, the word for *expected* [1].

Names that praise an individual or thing are also used, as seen in the Xhosa poet Bongani Sitole's tribute to Nelson Mandela.

> "Hail Dalibhunga!
> Words of truth have been exposed,
> He's the bull that kicks up dust and stones and breaks antheaps,
> He's the wild animal that stares at the sky
> Until the stars fall down" [1].

The names attached to HIV/AIDS bear similarities to the naming patterns used for leading personalities and mighty leaders.

Compare the praises of a great fighter, Mqikela Ndayi, a notable fighter, is called by a military metaphor: "He is a rifle speeding to its target." HIV/AIDS, a notable killer, has been given a similar name: "He who shoots to kill [1]."

A person living with AIDS is one who has suffered: *a hot coal fall upon himself or herself*. Such a person has *caught it* [1]. AIDS is *the One Who Finishes the Nation*, or *The Finisher of the Nation*. It is *The Killer of the Nation*. The names of AIDS are similar to those assigned to prominent Africans. The word for AIDS in Zulu and Xhosa is similar to that for the word plague. South Africa has eleven official languages. In English AIDS is *a disease for which there is no cure*; the word kill is not part of the description. The same is true in Afrikaans. Such is not the case with the nine other languages. Here, the Xhosa description is typical: "a disease which kills and which cannot be cured" [1]. Slaughtering of animals commonly accompanies funeral rites, and HIV/AIDS has been described as an illness that plagues the dwelling place of the living, *contaminating* a place such as KwaZulu–Natal, now styled KwaNyama–ayipheli, "at the place where the meat does not end." The reference to meat is not a symbol of prosperity but of the demand for meat at funerals. *Sodom and Gomorrah, a place from where you will not return* is another death-associated name for KwaZulu–Natal. *Kwatsi*, which translates as *disease* in the Sotho language, is linked in its provenance to a deadly disease that can spread from cattle to humans: anthrax. *Bacillus anthracis*, the scientific name for the anthrax organism, forms spores and can persist in very challenging conditions. It is highly resistant to cold and heath, to drying, and even to chemical disinfectants. Anthrax spores can live on and on in the soil [1,2]. The Sotho phrase *he or she is being held by HIV/AIDS* uses the word *kwatsi*, the anthrax-related word for disease. To be held by AIDS is, with gloomy symbolism, to be held by *kwatsi,* by anthrax [1].

23.2 Personification of AIDS

In African language, AIDS is personified as a killer: (1) *the beater-up of people*, (2) *the one who shoots to kill*, (3) *the finisher of the nation*, (4) *the indiscriminate killer*, and (5) *the one who chops down*. With a kind of gallows humor, other linguistic expression portrays it as a type of game, or as a train ride on an already overcrowded train. Zola and Umlazi are the names of places in Africa , the former located in the Soweto area, and the latter in KwaZulu–Natal. Thus, when it is said that the Zola or Umlazi railroad lines are overcrowded, the upshot is that these populous areas have high rates of human immunodeficiency virus (HIV) infection. Sometimes, the words for HIV or AIDS are simply avoided, as with a taboo. HIV/AIDS has been called, in reference to its being an acronym, *four words*, or *three words*. In reference to its seriousness, HIV/AIDS is *big words* or *the big matter*. The relationship between actual diagnosis and anxiety is also reflected in African naming patterns. Who is it that is afflicted with AIDS? A common response in the Zulu language runs contrary to accurate cause–effect logic but is, nonetheless, understandable: *People who go for tests* [1]. Like a person who is cheerful and hopeful until they find out from a doctor that they have cancer, so HIV-infected people often live without worrying. Following diagnosis that they are HIV seropositive, however, they then have both the infection and the worries that go with it in its diagnosed state.

Some Africans have adopted a conspiracy mentality with regards to AIDS. They accuse foreign nations of spreading AIDS to Africa: "Our leaders who travel to all these countries come with Aids." Race relations also influence the accusations: "Whites came to Africa with food that has Aids in it"; "The different nationalities that are filling up SA are those that are coming with these diseases" [1]. Many have also perceived a sinister foreign plot to end love among Africans, an attitude that implies racial mistrust as well.

Lottery imagery is used to refer to HIV/AIDS. The chances of winning the lottery are very small; those infected with HIV also have little hope of winning. *She is playing the lotto* is one allusion to an infected person; she or he *has the lotto* is another. Red-scarf imagery is used in reference to the symbolic AIDS ribbon or a red scarf: *She is wearing something round the neck* or *She is wearing the red scarf*. AIDS has also been called *the disease of beautiful people*. It brings great loneliness, and once who lives with AIDS is a *cow who eats alone*. That AIDS routinely brings life to a premature end is tragic enough, but this ultimate consequence is made sadder along the way by the effects of AIDS on personal relationships: "It ends love" [1].

References

1. Dowling T. Uqedisizwe—The Finisher of the Nation. HIV/AIDS and African Languages. 2006. http://www.africanvoices.co.za/research/aidsresearch.htm.
2. Stoltenow CL. Anthrax. 2000. http://www.ag.ndsu.edu/pubs/ansci/livestoc/a561w.htm, accessed December 21, 2011.

24
They Died of AIDS

HIESaren (AIDS in Basque)

24.1 AIDS attacks individuals

Berry Berenson died on September 11, 2001 (the September 11) aboard *American Airlines Flight 11*, the victim of the sinister terrorist plotting. Her husband, Anthony Perkins (1932–1992), died about a decade earlier of AIDS-related pneumonia. Perkins was closely associated with actor Rock Hudson (1925–1985), who acted in almost 70 movies before dying of an AIDS-related infirmity [1].

Keith Haring (1958–1990) painted a huge mural on the Berlin Wall at an internationally famous landmark, the Brandenburg Gate before he died of AIDS-related causes in 1990, and before that a number of his acquaintances also died of complications related to acquired immune deficiency syndrome (AIDS). (1). Africa is home to nearly 15% of the population of the planet, and in 2009, nearly three-fourths (72%) of all AIDS-related deaths occurred there [1].

Beauty and the Beast the movie featured an Academy Award winning song by the same name. Howard Ashman (1950–1991) wrote the lyrics before he died of AIDS-related causes. The Academy Award for the song was given to him in 1992, posthumously [1].

A child prodigy, the flamboyant piano genius Liberace (1919–1987) was the world's highest paid performer for a lengthy period of time. He first lost his health (which AIDS attacked), then his wealth (which death took from him) [1].

Born with an incredibly versatile voice with a vast range, Farrokh Bulsara, or Freddie Mercury (1946–1991) lived only 45 years; it was difficult to live longer with AIDS, a disease he disclosed the day before he died [2].

24.2 AIDS has a human face

After his eldest son, Makgatho Mandela (1951–2005), died of AIDS-related causes, AIDS activist and former South African president Nelson Rolihlahla Mandela (1918–2013) continued to lobby for AIDS awareness and funding. Mangosuthu Buthelezi (1928–), another noted South African political leader, lost two of his children to AIDS-related illnesses prior to the passing of Makgatho Mandela [3–5].

One of the most prolific authors imaginable, Isaac Asimov (1920–1992) wrote over 500 books before he died of AIDS-related causes. He might have written more had he lived free of human immunodeficiency virus (HIV) [6].

French interdisciplinary philosopher Michel Foucault (1926–1984) wrote of power and health policy, introduced new ways of thinking, and taught at the University of California at Berkeley before he died at the age of 57 of AIDS-related causes [7].

Hemophilia, a serious bleeding disorder, challenged Ryan White (1971–1990), but when a blood transfusion gave him HIV, his challenges became much more severe. Before dying of AIDS-related causes at the age of 18, Ryan challenged school authorities about his right to attend school, and helped educate numerous people about AIDS myths, and taught them about courage along the way [8].

After dying from AIDS-worsened pneumonia, his body lay in state at the Richmond, Virginia Governor's Mansion, an honor not given to anyone since Civil War General Thomas J. *Stonewall* Jackson's body lay there. Arthur Ashe (1943–1993) was a fabulous tennis player, and a caring human being who helped shatter AIDS stereotypes when he died [9].

AIDS is a human disease, and it does have a human face.

References

1. Johnson B. Top 10 Notable People Who Died from AIDS. http://list-verse.com/2011/12/01/top-10-notable-people-who-died-from-aids/, accessed December 14, 2011.
2. Gunn J and Jenkins J. Freddie Mercury: Biography. http://www.freddie.ru/e/bio/, accessed December 14, 2011.
3. Mandela's eldest son dies of Aids. *BBC News*. 6 January 2005. http://news.bbc.co.uk/2/hi/africa/4151159.stm, accessed December 14, 2011.

4. Mandela's son dies of AIDS at 54. *Reuters*. http://www.utexas.edu/conferences/africa/ads/233.html, accessed December 14, 2011.
5. Aids kills Zulu leader's daughter. *BBC News*. 7 August 2004. http://news.bbc.co.uk/2/hi/africa/3545450.stm, accessed December 15, 2011.
6. Isaac Asimov. http://www.asimovonline.com/asimov_home_page.html, accessed December 14, 2011.
7. Wohlsen M. Foucault at Berkeley: A university transformed. *Illuminations*. http://illuminations.berkeley.edu/archives/2005/history.php?volume=3, accessed December 21, 2011.
8. Johnson D. Ryan White Dies of AIDS at 18; His Struggle Helped Pierce Myths. Obituaries. *New York Times*. 9 April 1990. http://www.nytimes.com/1990/04/09/obituaries/ryan-white-dies-of-aids-at-18-his-struggle-helped-pierce-myths.html?pagewanted=all, accessed December 21, 2011.
9. Morrison A. Arthur Ashe. Biography. *CNNSI*. http://sportsillustrated.cnn.com/tennis/features/1997/arthurashe/biography1.html, accessed December 14, 2011.

25
Poetic Reflections on
the AIDS Crisis

SIDA (AIDS in Serbo–Croatian)

25.1 Etcetera: These are the houses that AIDS built

This is the coffin that gave the husband who succumbed to acquired immune deficiency syndrome (AIDS) a new-yet-smaller wooden residence.

This is the tree that supplied the wood to build for the husband who succumbed to AIDS a new-yet-smaller wooden residence.

This is the son who chopped down the tree that supplied the wood to build for the husband who succumbed to AIDS a new-yet-smaller wooden residence.

This is the human immunodeficiency virus-infected (HIV-infected), grieving spouse who gave birth to the son who chopped down the tree that supplied the wood to build for the husband who succumbed to AIDS a new-yet-smaller wooden residence.

This is the tree-chopping orphan of the now deceased HIV-infected, grieving spouse who gave birth to the son who chopped down the tree that supplied the wood to build for the husband who succumbed to AIDS a new-yet-smaller residence.

Etcetera.

P.S: This is HAART, but there is never enough, and it only delays, never cures.

25.2 AIDS personified

AIDS ignores political boundaries, crosses porous borders of poor human choice, pays no customs duties, travels with neither passport nor visa, drives the world's highways, sails the planet's seas, flies through earth's airspace, breaks hearts, breaks down immunity, covers truth, spreads myths, promotes ignorance, wrecks homes, grieves children, fathers orphans, leaves widows, chops trees for coffins, exploits poor choices, gnaws away at health, robs the rich, impoverishes the poor, destroys CD4+ counts, baffles scientists, challenges budgets, reduces profits, encourages absenteeism, promotes burials, worships immorality, upholds double standards, exploits women, exploits men, exploits children, flourishes in an apathetic NIMBY (not in my backyard) environment, avoids unified NIOBY (not in our backyard) efforts [1], evades classical immunity, avoids slumber, takes no vacations, torments the adults in Cape Town and Mumbai, confuses the children in orphanages and on streets, dries no tears, sheds no tears, distributes no handkerchiefs, makes no mortgage payments, slays mortgage payers, robs down payments, kills down payers, evicts tenants, plays Trojan horse, shares freely, terminates employment, drives up insurance rates, drives down population growth, loves drug users, mixes well with alcohol, encourages fear, squelches hope, creates morbid personal histories, changes the history of nations, tortures the present, and slays the future.

Reference

1. Pace DG. From Nimby problems to Nioby complexes: Wicked webs as international policy dilemmas. In Phelps GA, ed., *Cross Currents: Renewable Energy Use, 1997–1998*, pp. 5–39, Clayton, MO: The Alan Company, 1997.

Index

Note: Page numbers followed by f refer to figures.

POCKET GUIDES TO
BIOMEDICAL SCIENCES

Series Editor
Dongyou Liu

A Guide to AIDS
Omar Bagasra and Donald Gene Pace

Tumors & Cancers: Brain – Central Nervous System
Dongyou Liu

A Guide to Bioethics
Emmanuel A. Kornyo

Tumors & Cancers: Head – Neck – Heart – Lung – Gut
Dongyou Liu